Advanced Materials for Sodium Ion Storage

Advanced Materials for Sodium Ion Storage

Ranjusha Rajagopalan

Lei Zhang

CRC Press
Taylor & Francis Group
Boca Raton London New York

CRC Press is an imprint of the
Taylor & Francis Group, an **informa** business

CRC Press
Taylor & Francis Group
52 Vanderbilt Avenue,
New York, NY 10017

© 2020 by Taylor & Francis Group, LLC
CRC Press is an imprint of Taylor & Francis Group, an Informa business

No claim to original U.S. Government works

Printed on acid-free paper

International Standard Book Number-13: 978-1-138-38965-6 (Hardback)

This book contains information obtained from authentic and highly regarded sources. Reasonable efforts have been made to publish reliable data and information, but the author and publisher cannot assume responsibility for the validity of all materials or the consequences of their use. The authors and publishers have attempted to trace the copyright holders of all material reproduced in this publication and apologize to copyright holders if permission to publish in this form has not been obtained. If any copyright material has not been acknowledged please write and let us know so we may rectify in any future reprint.

Except as permitted under U.S. Copyright Law, no part of this book may be reprinted, reproduced, transmitted, or utilized in any form by any electronic, mechanical, or other means, now known or hereafter invented, including photocopying, microfilming, and recording, or in any information storage or retrieval system, without written permission from the publishers.

For permission to photocopy or use material electronically from this work, please access www.copyright.com (http://www.copyright.com/) or contact the Copyright Clearance Center, Inc. (CCC), 222 Rosewood Drive, Danvers, MA 01923, 978-750-8400. CCC is a not-for-profit organization that provides licenses and registration for a variety of users. For organizations that have been granted a photocopy license by the CCC, a separate system of payment has been arranged.

Trademark Notice: Product or corporate names may be trademarks or registered trademarks, and are used only for identification and explanation without intent to infringe.

Library of Congress Cataloging-in-Publication Data

Names: Rajagopalan, Ranjusha, author. | Zhang, Lei (Researcher in energy storage materials), author.
Title: Advanced materials for sodium ion storage / Ranjusha Rajagopalan and Lei Zhang.
Description: New York, NY : CRC Press/Taylor & Francis Group, 2020. | Includes bibliographical references and index.
Identifiers: LCCN 2019014827 | ISBN 9781138389656 (hardback : acid-free paper) | ISBN 9780429423772 (ebook)
Subjects: LCSH: Sodium ion batteries.
Classification: LCC TK2945.S62 R34 2020 | DDC 621.31/242--dc23
LC record available at https://lccn.loc.gov/2019014827

Visit the Taylor & Francis Web site at
http://www.taylorandfrancis.com

and the CRC Press Web site at
http://www.crcpress.com

Contents

Preface ix
Acknowledgments xi
Authors xiii

1. Introduction for Sodium Ion Batteries 1
   1.1   The History of the Sodium Ion Battery 1
   1.2   The Main Challenges for the Sodium Ion Batteries 3
         1.2.1   Cathode Materials 3
         1.2.2   Anode Materials 4
         1.2.3   Electrolyte and Additions 5
   Reference 6

2. Electrochemical Reaction Mechanism in Sodium Ion Batteries 7
   2.1   Introduction 7
   2.2   Key Components and Common Terminologies 9
         2.2.1   Key Components 9
         2.2.2   Key Components Selection Criteria 10
         2.2.3   Common Terminologies 10
                2.2.3.1   Battery Basics 10
                2.2.3.2   Variables to Describe the Battery Condition 12
                2.2.3.3   Battery Technical Specifications 16
   2.3   Electrochemical Reaction Mechanism 18
   References 20

3. Anode Materials for Sodium Ion Batteries 23
   3.1   Introduction 23
   3.2   Insertion Materials 23
         3.2.1   Carbon-Based Anode Materials 24
                3.2.1.1   Graphite 24
                3.2.1.2   Hard Carbon 26
                3.2.1.3   Graphene 29
                3.2.1.4   Heteroatom Doping 30
                3.2.1.5   Biomass-Derived Carbon 31
         3.2.2   Titanium-Based Materials 32
                3.2.2.1 TiO_2 33
                3.2.2.2   Lithium Titanate 37
                3.2.2.3 $Na_2Ti_3O_7$ 40
   3.3   Conversion Electrode Materials 41
         3.3.1   Transition Metal Oxides 42
                3.3.1.1   Iron Oxides 42

			3.3.1.2	Cobalt Oxides 43
			3.3.1.3	Tin Oxides 44
			3.3.1.4	Copper Oxides 45
		3.3.2	Transition Metal Sulfide 45
			3.3.2.1	CoS_x (x = 1 and 2) 46
			3.3.2.2	MoS_2 47
			3.3.2.3	FeS_2 47
			3.3.2.4	SnS_x (x = 1 or 2) 48
		3.3.3	Transition Metal Phosphide 48
	3.4	Alloying Reaction Materials 49
		3.4.1	Silicon, Germanium, Tin 50
			3.4.1.1	Si 50
			3.4.1.2	Ge 52
			3.4.1.3	Sn 52
		3.4.2	Antimony, Phosphorus, Bismuth, and Arsenic 54
			3.4.2.1	Sb 54
			3.4.2.2	P 57
		3.4.3	Binary Inter-metallic Compounds 60
	3.5	Organic Compounds 62
References 66

4. Cathode Materials for Sodium Ion Batteries 83
4.1	Introduction 83
4.2	Transition Metal Oxides 84
	4.2.1	Layered Sodium Metal Oxides 84
		4.2.1.1 Sodium-Based Single Transition-Metal Oxides 84
		4.2.1.2 Sodium-Based Mixed-Cation Oxides 88
	4.2.2	Sodium-Free Metal Oxides 90
		4.2.2.1 Vanadium Oxides 90
		4.2.2.2 Manganese Oxides 91
		4.2.2.3 Tunnel-Type Sodium Transition-Metal Oxides 92
4.3	Transition-Metal Fluorides 93
	4.3.1	Na-Free Transition-Metal Fluorides 94
	4.3.2	Na-Based Transition-Metal Fluorides 94
4.4	Polyanion Compounds 95
	4.4.1	Phosphates 95
		4.4.1.1 Olivine 95
		4.4.1.2 Sodium Super Ionic Conductor (NASICON) 98
		4.4.1.3 Vanadyl Phosphate 100
		4.4.1.4 Pyrophosphates 100
		4.4.1.5 Fluorophosphates 101
		4.4.1.6 Mixed Polyanion 103
		4.4.1.7 Alluaudites 105

		4.4.2	Hexacyanoferrates 105
		4.4.3	Sulfates 107
	4.5	Organic Materials 108
		4.5.1	Non-polymeric Materials 109
		4.5.2	Polymeric Materials 111
	4.6	Conclusions and Future Outlook 112
	References 115

5. Electrolytes, Additives, and Binders for Sodium Ion Batteries 131
    5.1 Introduction 131
    5.2 Electrolytes 132
    5.3 Additives 135
    5.4 Binders 137
    References 140

6. Current Challenges and Future Perspectives 143
    6.1 Introduction 143
    6.2 Current Challenges 144
        6.2.1 Electrode Selection 144
        6.2.2 Electrolyte, Additives, and Binder Selection 146
        6.2.3 Full-Cell Performance 147
    6.3 Future Perspectives 148
    References 150

Index 153

Preface

The global demand for electricity is huge, and it's growing by ~3.6% annually. Resources are running low, pollution is increasing, and the climate is changing. As we are about to run out of fossil fuels in the next few decades, we are keen to find substitutes that will guarantee our sustenance on energy on a long-term basis. With the growing importance of renewable energy sources, there has been a quest to enhance efficiencies and to lower the costs of these technologies. Modern technology is already providing us with such alternatives like wind turbines, photovoltaic cells, biomass plants, and more. However, compared to traditional power plants, they produce much smaller amounts of electricity and even more problematic is the inconsistency of the production. Storing energy is not an easy task. Our smartphone battery only lasts for about a day, while laptops last only a few hours. The range for electric cars is limited to 100 km. While these are only examples for comparatively small devices, the problem escalates when storing energy at the level of hundreds to thousands of wind turbines and photovoltaic cells. For technical reasons, however, the amount of electricity fed into the power grid must always remain on the same level as demanded by the consumers to prevent blackouts and damage to the grid. It leads to situations where the production is higher than the consumption or vice versa. This is where storage technologies like batteries come into play, as they are the key element to balance out these flaws. Among batteries, lithium ion batteries (LIBs) have dominated the field of energy storage systems in the commercial market, particularly in the area of portable devices, such as electronics and communications gadgets. However, due to the limited resources of lithium in the earth, the price of the Li-based electrode materials has been significantly increased during the past few years. Replacing LIB with sodium ion battery (SIB) for practical application is crucial for future development, due to the practically "unlimited" nature of sodium. The development of SIBs is mainly based on the designing of novel electrode materials, utilization of appropriate electrolytes and additives, along with the improvement of the cell fabrication techniques. Materials scientists and electrochemists are constantly trying to push the for commercialization of SIBs. Thus, promising electrodes with lower cost and good electrochemical performance in SIBs are on the horizon.

This book provides the cutting-edge knowledge and novel concepts of the advanced SIBs. The chapters have been written in a manner that fits the background of different science and engineering fields. Also, the complete set of topics contained in this book can be covered in a single semester and prepare the student for a research program in the advancing field of batteries, apart from equipping the student for mastering the subject.

Chapter 1 gives a general review about the history of the sodium ion batteries' development. An overview of electrodes, electrolytes, and additives, is also provided in Chapter 1. Chapter 2 provides the detailed discussion of the reaction mechanism of sodium ion batteries. A thorough understanding of the key components and the fundamental electrochemical mechanism which governs the storage property are also summarized. The main objective of this chapter is to understand the underlying charge storage mechanism in the SIBs. Chapter 3 listed all the promising anode materials. The main disadvantages and advantages of these anode electrodes are compared, and the approaches about how to improve the electrochemical performance of the anodes are provided. The future developmental directions of the anode materials, such as developing novel organic and alloying-typed anode materials, are highly recommended in this chapter. Chapter 4 is a general overview of the development of the cathode materials in sodium ion batteries. The ones which are attracting the most research attention are discussed and summarized. In this chapter, through the in-depth understanding of the promising cathode candidates, the causes behind the drawbacks of the existing cathode systems, and how these issues can be addressed by designing new cathodes with improved electrochemical performance are revealed and explained. In addition, the challenges and future prospects are also incorporated to provide insight and guidance to the readers for designing and developing new cathode systems with enhanced electrochemical properties for SIBs. Chapter 5 provides an overview about the currently commercialized electrolytes, binders, and additives in the SIBs. The potential advanced binders and additives are also reviewed. In the last chapter (Chapter 6), in order to bring this potential system to the market, the underlying problems and the potential advantages of the current SIBs are concluded. This chapter discusses the current challenges in various fields and recommends some future directions for SIBs; this could provide significant insights into fundamental and practical issues in the progress of SIBs.

Dr. Ranjusha Rajagopalan
Hunan Provincial Key Laboratory of Chemical Power Sources
College of Chemistry and Chemical Engineering
Central South University
Changsha, P.R. China

Dr. Lei Zhang
Centre for Clean Environment and Energy
Environmental Futures Research Institute
Griffith University
Gold Coast, Australia

Acknowledgments

The authors acknowledge Prof. Shi Xue Dou, Prof. Hua Kun Liu, Prof. Huijun Zhao, Prof. Yougen Tang, and Prof. Haiyan Wang for their support. National Nature Science Foundation of China (No. 21671200 and No. 21571189), Hunan Provincial Natural Science Foundation of China (No.2018JJ4002), Hunan Provincial Science and Technology Major Project of China (No.2017GK1040), and Hunan Provincial Science and Technology Plan Project, China (No. 2017TP1001) are acknowledged for the financial support. The Postdoctoral International Exchange Program Project (No. [2018]115), China, is also greatly acknowledged for the financial support. Further, China Postdoctoral Science Foundation (2019M652802) is also greatly acknowledged for the financial support. Finally, the authors thank all the family members and friends for their continuous support during the preparation of this book.

Authors

Dr. Ranjusha Rajagopalan earned her PhD in Physics with specialization in battery technology from the University of Wollongong, Australia in 2017. She is currently working on functionalized organic/inorganic materials for developing high-performance metal ion batteries, including sodium and potassium ion batteries at College of Chemistry and Chemical Engineering, Central South University, China. She obtained her master's and bachelor's degrees in physics from Calicut University, India.

Dr. Rajagopalan's research interests include enhancement of power and cycling performances of battery electrodes by designing novel materials and composites. As part of her collaborative research and exchange programs, she has worked in many foreign universities like Nanyang Technological University, Singapore; Sichuan University, China; and Pai Chai University, S. Korea. She is a recipient of several prestigious fellowships/funding, which include, Postdoctoral International Exchange Program Funding, China 2018–2020; China Postdoctoral Science Foundation Funding 2019 Faculty Scholarship (Research and Innovation division) UOW, Australia, 2014; CSIR (Council of Scientific and Industrial Research) Scholarship, India, 2012; MNRE (Ministry of New and Renewable Energy) Scholarship, India, 2011–2012; and DST (Department of Science and Technology) Scholarship, India, 2010–2011. RRC (Resources Regional Center) Scholarship, South Korea, 2011. She has published over 34 journal research articles, 2 book chapters, and 1 book and has filed 5 Indian patents in the areas of energy storage and generations.

Dr. Lei Zhang earned his PhD in Physics with specialization in energy storage systems from University of Wollongong, Australia, in the year 2017. He is currently working on symmetric electrodes for developing novel energy storage systems at Centre for Clean Environment and Energy, Griffith University, Australia. He received his master's degree from Qingdao University Science and Technology, China.

His research mainly focuses on the improvement of cycling performances of silicon-based lithium ion batteries by the introduction of the porous structure and conductive coatings. He has published over 25 research articles as the first or corresponding author. He is also the reviewer of relevant journals, including the *ACS Nano*, *Nano Energy*, and *ACS Applied Materials & Interfaces*.

1
Introduction for Sodium Ion Batteries

The depletion of fossil fuels increases the concern about environmental pollution. As a result, tremendous attention has been paid to novel renewable and cleaner energy sources, including wave, wind, and solar. Producing low cost and green renewable energy from these sources plays an important role in satisfying the growing demand for transportation fuel, heat, and electricity. Unfortunately, the intermittent ability of the renewable primary energy limits its effective practical application, which will result in instability of the power supply. From this point of view, the energy storage systems (EESs) are the most crucial aspects during the renewable energy production process, because the renewable energy can be stored gradually without concerns about its intermittent ability. According to the current understanding for energy production, it can be produced and stored with the help of different means, such as electrical, mechanical, electrochemical, or chemical means. Among these four different EESs technologies, it is promising to take advantage of the electrochemical secondary battery for large-scale storage of electricity because of its high flexibility to different conditions, relevant high energy conversion efficiency, and facial maintenance.

1.1 The History of the Sodium Ion Battery

The first commercialized electrochemical-based EES was the lithium ion battery (LIB) which was initially commercialized by Sony in 1991. LIBs quickly dominated the market of portable electronic devices and were identified as the most promising candidate to address the electrical grid concerns due to the high-specific storage capacity, long cycling life, and suitable working potential. However, with the increasing lithium demand within the last 10 years, the price of lithium compound is increased significantly due to the limited global lithium reserves. Worldwide concerns and tremendous attention have been paid to search for new electrochemical-based EESs.

Compared with LIBs, sodium ion batteries (SIBs) and potassium ion batteries (KIBs) have attracted great efforts in the past few years due to the practically unlimited nature of sodium and potassium resources. Unfortunately, potassium shows the largest ionic radius (~0.4 Å larger than Na$^+$ and ~0.7 Å larger than Li$^+$), leading to the limited choice of the available anodes and poor cycling stability of the electrodes in KIBs. Compared with Li and K, Na is the most abundant element on earth. In addition, a great number of Na-containing resources is available in the commercial market, which provides the opportunity for scalable electrode preparation. For example, 23 billion tons of soda ash is available in America alone. As a result, the cost for the SIBs electrodes are largely reduced when compared with LIBs, which makes SIBs more promising as the next generation EES and attracted tremendous research attention during the past few years.

Actually, the initial study of SIBs was started from the 1970s which is quite close to the first research about LIBs. Unfortunately, the following research on SIBs was abandoned after that because of the significant breakthrough in the successful commercialization of LIBs. With the rapid development of LIBs, the high cost and limited resource of Li contributed the transfer of the research focus from LIBs to SIBs. Exploring and developing effective electrodes and electrolytes for the SIBs now represent a hot research topic. Figure 1.1 shows the recent progress in SIBs.[1]

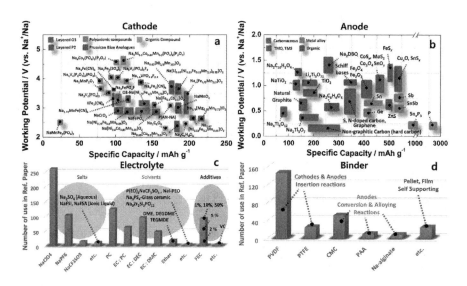

FIGURE 1.1
Recent research progress in sodium ion batteries: (a) cathode, (b) anode, (c) electrolyte, and (d) binder. (Copyright 2017, The Royal Society of Chemistry.)

1.2 The Main Challenges for the Sodium Ion Batteries

SIBs and LIBs have similar components inside the systems, such as the anode, cathode, electrolyte, and separator (as shown in Figure 1.2).[1] Furthermore, there is no significant electrical storage mechanism difference between SIBs and LIBs except for their ion carriers, which enable most of the electrode materials directly used from LIBs to SIBs. However, some differences still exist between LIBs and SIBs. For example, Na$^+$ ions are 0.26 Å larger than the Li$^+$ ions, leading to the relevant differences in the solid-electrolyte-interfaces (SEIs) formation, transport ability, and phase stability. In addition, Na$^+$ ions have a higher atom mass and standard electrode potential (−2.71 V vs SHE for Na$^+$ as compared to −3.02 V vs Standard Hydrogen Electrode (SHE) for Li$^+$) than the Li$^+$, leading to the lower energy density for SIBs.

1.2.1 Cathode Materials

According to the current development of SIBs, the overall ability of SIBs is determined by the performance of the cathode materials. Therefore, it is important to develop promising cathode materials with outstanding electrochemical abilities. Unfortunately, most attempts to further improve the cathodes have tackled the problem at the poor rate performance and the low specific capacity. For example, sluggish kinetics during sodiation/desodiation process and difficult transportation of the Na ions across the

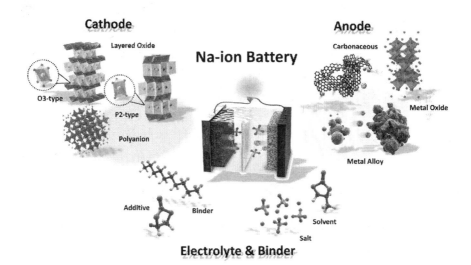

FIGURE 1.2
Illustration of a Na-ion battery system. (Copyright 2017, The Royal Society of Chemistry.)

host-material framework happened due to the larger Na⁺ radius (0.98 Å) than Li⁺ (0.69 Å), resulting in the degradation in specific capacity and rate capacity. In addition, larger volume changes of the active materials will cause phase changes, resulting in deterioration in the cycling performance. More importantly, the low specific capacity of the most currently used cathodes limit the practical application of the full cells. Therefore, many recent research works have focused on efforts to overcome all these disadvantages by exploring and exploiting the potential properties of sodium-based cathodes.

Four different kinds of materials can be used as promising cathode materials for SIBs, including transition materials, transition-metal fluorides, polyanion compounds, and organic materials. The transition materials, which mainly include layered sodium metal oxides and sodium-free metal oxides, were extensively studied during the 1970s and 1980s. The common molecule formula for these kinds of sodium metal oxides can be described as $Na_xMO_{2y}^+$, where M is a transition metal. Apart from the transition materials, two main fluorides, such as the Perovskite transition-metal fluorides MF_3 and sodium fluoroperovskites $NaMF_3$ (M = Ni, Fe, Mn), can be treated as cathodes for SIBs. The stronger ionic bonding and electronegativity of fluorine than oxygen made these fluoride-based cathodes potential candidates for SIBs; especially due to their high operating voltages. Polyanionic phosphate compounds have been successfully used as cathode electrodes in LIBs. Therefore, it is interesting to explore the electrochemical ability of these polyanion phosphate and mixed polyanion materials as a cathode system in SIBs. After the relevant research during the past few years, these polyanionic systems are regarded as potential candidates for the practical SIBs due to their good structural stability, diversity, superior operating potential, thermal safety, high cycling life, and strong inductive phenomenon resulting from the highly electronegative anions. Different from the abovementioned inorganic materials, the organic compounds show unique abilities when used as cathodes in SIBs. For example, the organic electrodes are endowed with various advantages, such as structural diversity, less expensive, good safety, recyclable nature, and flexibility. Generally, two different kinds of organic materials are attracting the most research attention, these are metal organic materials and metal-free organic materials.

1.2.2 Anode Materials

Compared with the relevant research about cathode materials in SIBs, it is more difficult to find the appropriate anode candidates in SIBs. This is mainly because limited materials have been reported to be available as negative electrodes in SIBs. It can be seen that most of the previous reported cathode materials for LIBs can be used as cathode materials in SIBs just by replacing the lithium with sodium. Unfortunately, based on the currently promising

Introduction for Sodium Ion Batteries

anodes, such as graphite, which is a traditional commercial anode material in LIBs, most of them cannot be directly used as anodes in SIBs because SEI film generated on the surface of the active particles is unstable. As a result, the exploration for novel commercial anodes in SIBs is very important.

The current research about the anode materials for SIBs can be divided into four different categories based on the different electrochemical reaction mechanism between the anode materials and reversible Na^+: the insertion materials, conversion materials, alloying materials, and organic materials. The carbon- and titanium-based materials are the most popular ones in the insertion materials, because the Na ions can be reversibly inserted/extracted into/from these kinds of materials due to their unique structural abilities. Conversion-type material, such as the transition metal sulfide, transition metal oxide, and transition metal phosphide, is another promising candidate for reversible energy storage, because it has high theoretical specific capacity for adopting Na^+ during the conversion process. For the conversion materials, a new compound will be generated after the sodiation process, leading to the high reversible specific capacity. Alloying-type materials show promising electrochemical performance when they are employed as anodes for SIBs, because the alloying-type materials provide abundant active positions for Na ion insertion/extraction under a very low working voltage range. Various kinds of interactions can be conducted with Na ions, such as Sn, Bi, Si, Ge, As, Sb, and P. Compared with the abovementioned inorganic electrode materials, the organic ones give energy storage a new lease on life and provide benefits in terms of their mechanical flexibility, large stable structure changes, low density, low cost, environmentally friendly ability, and chemical diversity, which provide a promising application of organic materials as candidate for special batteries.

1.2.3 Electrolyte and Additions

As for the electrolyte and other additions, even more limited work has been conducted or published when compared with cathodes and anodes. The main reason for that is because the research work on the electrolyte and additions is more likely an interdisciplinary project which combines the research cooperation from the organic materials, inorganic materials, and electrochemical materials. As a result, poor research outcomes on these areas have been obtained during the past few years.

The practical application of the commercial full batteries is usually hindered by the decomposition of the liquid electrolytes under higher working voltage, leading to the lower energy densities of the full cells. Apart from that, the important SEI films are generated during the charging/discharging process. And the structural stability of this SEI film is mainly determined by the ingredients of the electrolytes, additions, and binders. Therefore, it is important to develop novel electrolytes and additions to satisfy the requirements from the future energy storage systems.

Some necessary requirements and key points of the electrolyte are needed for the practical application of the SIBs, including the chemical inertness, low cost, scalable production available, electrical insulating and ionically conductive, environmentally friendly, and facial fabrication process. For the liquid electrolyte, the polar ability should be well maintained even under high dielectric constant. In addition, it should have a low viscosity to enhance the transfer of ions. As for the additive inside the electrolyte, it is required to generate the stable SEI films between the electrolyte and electrode. Moreover, the safety of the full cells and the electrochemical stability of the electrolytes should be improved with the help of using extra additives. But for the binders, the most important role is to maintain the structural stability of the electrodes during the cycling process. For example, because of the large volume changes of most anode materials, a suitable binder could effectively address this problem via the strong chemical bonding between binder and electrode, leading to improved structural stability. Furthermore, a tight contact between the electrode materials and the current collector can also be obtained with the introduction of the binders, which can significantly avoid the stripping of the active materials from the current collector.

In summary, it can found that the research about the SIBs is relatively limited and more efforts need to be made to further improve its electrochemical ability. Therefore, in this book, we will emphasize the current research development of the cathodes, anodes, and electrolytes of the SIBs to give an overall understanding for the SIBs.

Reference

1. Hwang, J. Y., Myung, S. T., Sun, Y. K. Sodium-ion batteries: Present and future. *Chem. Soc. Rev.* **2017**, *46* (12), 3529–3614.

2

Electrochemical Reaction Mechanism in Sodium Ion Batteries

2.1 Introduction

The modern world is getting more and more dependent on energy to have an uninterrupted power supply for the continuous usage of Internet and electronic gadgets. However, in recent years the energy supply could not meet the high energy demand, leading to the energy crisis. The Industrial Revolution, one of the reasons for this disparity, tremendously altered social life, technologies, transportation, and production activities. Fossil fuels, for instance, coal, crude oil, and natural gas, are utilized as the prime energy sources to supply power for most of the energy required for human activities.[1] But these power productions resulted in many adverse effects on the environment and human health.[2,3] Moreover, the continuous usage of the energy led to the depletion of the natural reserves of these fossil fuels. Thus, the advancement of safe, environmentally friendly, low cost, and sustainable energy technologies is one of the prime challenges in the current energy research sector. This green energy technology can be divided into three main parts: (1) clean energy generation (e.g., solar energy, wind, hydropower, biomass plants, geothermal, etc.); (2) energy storage (e.g., batteries, hydrogen, biofuels, supercapacitors, etc.); and (3) energy management and efficient utilization of energy (e.g., smart buildings and efficient lighting systems).[4-6] Among the different energy storage technologies, lithium ion batteries (LIBs) are considered to be an effective and efficient energy storage unit for a variety of portable applications (e.g., cellular phones, laptop computers, digital cameras, etc.).[7-9] Prior to the commercialization of LIBs, the Ni/Cd battery had been the only suitable energy storage unit for different portable applications, such as wireless communications, mobile computing, etc. In 1991, Sony and Asahi Kasei companies released the first commercial LIB onto the market. Later in 2006, commercialization of LiFePO$_4$ had led a different dimension in LIBs technologies for portable electronic gadgets. As of 2018, LIBs are the leader among rechargeable batteries for portable gadgets/electric vehicle applications. Durability, power, and energy performances are the three principal factors for any energy storage units, which make the technology promising for any portable electronic gadgets and electric vehicle applications. As mentioned above, at present the LIBs are considered to have all these properties and thus are appropriate for different energy applications. However, the growing demands for these LIB systems are yet to be complemented by supply. Consequently, in the future it will be impossible for stand-alone lithium-based devises to satisfy the ever-increasing demands for grid-scale energy storage, portable electronic gadgets, and automobile applications. Moreover, since LIB applications move to grid-scale and automobile from simple portable electronic; the cost of the lithium-based materials has been increased tremendously.[10] The development of LIBs, consequently, depended upon the market price of lithium resources. Thus, replacing the LIBs with cost-effective and abundant alternatives could leave the batteries' future less susceptible to price variations when the market expands.[11] These concerns and huge demands have made researchers focus on alternative energy storage devices like sodium ion batteries (SIBs), potassium ion batteries, aluminum ion batteries, and magnesium ion batteries. Among these battery technologies, SIBs are considered to be an appropriate alternative to LIBs, since the elemental properties and electrochemical storage process of lithium and sodium are similar (Table 2.1). Other notable advantages of sodium are its abundance (nearly inexhaustible) and its uniform distribution around the globe having a Clarke's number of 2.644 (measure of material's abundance), which makes the sodium-based materials more economical and thus could meet the increasing energy demands in the future. Nonetheless, by taking their energy density and durability into consideration, the SIBs are not yet up to the LIBs. In order to address these issues, new high performing and stable cathode/anode materials need to be investigated. Therefore, to explore and exploit new materials, first a thorough understanding of the key components and the fundamental electrochemical mechanism which governs the storage property needs to be unraveled. Thus, the objective of this chapter is mainly to understand the underlying charge storage mechanism in the SIBs.

TABLE 2.1

Property Comparison of Lithium and Sodium

	Na	Li
Atomic radius (pm)	190	167
Ionic radius (pm)	102	76
Atomic weight (g mol^{-1})	22.989	6.94
E_0 vs. SHE (V)	−2.7	−3.04
Melting point (°C)	97.79	180.54
Boiling point (°C)	882.94	1347
Crystal structure	Cubic	Cubic
Density at 293 K (g cm^{-3})	0.971	0.53
Classification	Alkali metal	Alkali metal
Clarke's number	2.644	<0.05
Distribution	Everywhere	70% in South America
Price/kg, carbonate ($)	0.3	6.1

2.2 Key Components and Common Terminologies

To design and understand the electrochemical mechanism of a battery, it is important to know the key components that govern its charge storage mechanisms. The battery design consists of a combination of two or more cells connected in a series/parallel mode to meet the specific requirements. The cell is the basic electrochemical entity that includes all the below mentioned components to generate electrical energy from the stored chemical energy. The different components and important terminologies in a cell can be defined as follows.

2.2.1 Key Components

The SIBs consist of three main components: electrodes (cathode and anode), electrolyte, and separator. Each of these components is crucial for the proper functioning of the electrochemical cell.

Electrodes: The electrodes are the basic units in an electrochemical cell. The two electrodes in an electrochemical cell are assigned as positive (in general cathode) and negative (in general anode). In other words, a cathode is an oxidizing electrode that gets electrons from the external circuit and thus undergoes reduction during the electrochemical process. The cathode is comprised of active material (the main cathode material which determines the capacity and voltage of the cell), conductive additive (generally carbon black or other conducting carbon), and binder (adhesive which keeps the active material and the conductive additive firmly settled on the aluminum substrate) coated onto the thin aluminum substrate. On the other hand, an anode is the reducing electrode that gives electrons to the external circuit and gets oxidized during the electrochemical process. Like cathode, anode also contains active materials, conductive additives, and binders coated onto thin copper substrate (unlike LIBs, SIBs can use aluminum substrate instead of copper on the anode side, which help to reduce the cost of the SIBs). The voltage of a cell is the potential difference between the anode and cathode.

Electrolyte: A medium that allows the ions to travel between the electrodes (anode and cathode). In general, the electrolyte will be in liquid form, where the acid or alkali salt dissolved in water, organic, or other solvents. The additives can be also added to the electrolyte for the specific purpose. Apart from the liquid electrolytes, it should be also noted that the researchers are working on different solid-state electrolytes to develop safe SIBs. For both liquid and solid-state electrolytes, the materials used to prepare the electrolyte should provide good ionic conductivity to ensure smooth sodium ion movements across the electrodes.

Separator: A porous membrane which avoids the short circuiting of the cathode and the anode, thus, to ensure the safety of the cell, whereas, it makes sure of the free motion of ions through its pores. In a way, it is an electron insulator, but an ion conductor.

2.2.2 Key Components Selection Criteria

The proper combinations of anode, cathode, and electrolyte materials could result in a cell which is having all the desired properties, such as light-weight, high voltage, superior capacity, and better power performances. However, such combinations may not be practical because of a number of reasons, like material/processing cost, handling/fabrication difficulties, etc. Nevertheless, it should be useful to have some of the desirable properties while selecting the anode, cathode, and electrolyte materials for the cell fabrications. Table 2.2 provides some desired properties of battery components, which could be useful while selecting the components.

2.2.3 Common Terminologies

2.2.3.1 Battery Basics

Cell, Battery, and Packs: A cell is the smallest and basic electrochemical unit, which consists of key electrochemical components such as anode, cathode, electrolyte, separator, etc. There are different forms of cells such as coin, prismatic, cylindrical, pouch, etc. Figure 2.1 demonstrates the different cell formats.[12] A battery is the combination of two or more electrochemical cells connected in series or parallel manner in order to obtain the desired specifications. A battery pack is the encased unit which consists of a set of batteries or electrochemical cells assembled in series and/or parallel manner along with

TABLE 2.2

Desired Properties of Battery Components

Cathode	Anode	Electrolyte
Non-reactive with electrolyte	Non-reactive with electrolyte	Non-reactive with electrodes
Good oxidizing agent	Good reducing agent	Should not conduct electricity
Appropriate operating voltage	High Coulumbic efficiency	Superior ionic conductivity
Structural stability	Structural stability	Safe to handle
Cost effective	Cost effective	Cost effective
Ease of synthesis	Ease of synthesis	Properties should not change with temperature variations
Good conductivity	Good conductivity	

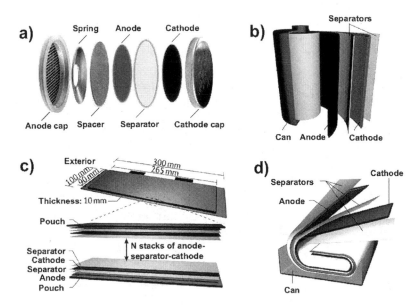

FIGURE 2.1
(a) Coin cell, (b) cylindrical cell, (c) pouch cell, and (d) prismatic cell format (Figure (b–d) Reprinted with permission from *Nat. Rev. Mater.*, 2016, 1, 16013. Copyright (2016) Springer Nature).

sensors, regulators, balancers, controllers (e.g., battery management systems [BMS] and thermal management systems [TMS]), etc., to provide the needed energy, power, and voltage.

For large-scale applications, for instance, in electric vehicles, a particular number of cells are assembled into a module according to the requirement. In general, the design of such a module depends mainly on the size/shape of the products, the interconnecting circuits, and different safety, temperature controllers. The larger commercial cell formats to design modules for different applications are cylindrical, prismatic, and pouch (Figure 2.1b–d). Cylindrical cell formats in most applications, including Tesla Motors' vehicles modules, follow a standard size model of 18650 with diameter and length of 18 and 65 mm, respectively. For a typical cylindrical cell format, the volumetric energy density will be ~20% higher than that of the pouch and prismatic cell formats, due to the higher tension wounding of the stacked cells.[13,14] However, in spite of the higher energy densities of cylindrical cells, pouch and prismatic cells are adopted for various applications because of their lesser dead volumes on the module level and superior degrees of design freedom. Apart from that, unlike cylindrical cells, the size of pouch and prismatic cells can be easily customized according to the requirements. For instance, the pouch cell shown in Figure 2.1c, demonstrates a dimension of 300 × 100 × 10 mm (length × width × thickness), which is a typical product of one of the battery manufacturers (SK Innovation). Inside this cell prototype, n number of anode-separator-cathode units are stacked to achieve the given thickness and specification, with both surfaces of each current collector (Al and Cu foil), apart from the outermost stacks, coated with electrode (cathode and anode) materials to achieve high volumetric energy out of the module.

Battery classifications according to the performance: The main two battery requirements are power and energy. Thus, batteries are generally prepared either to provide high power or high energy according to the requirements. Thus, the main two classifications are based on its power or energy performances. Another common classification is the high durability; this category provides better cycle life.

C-rate and E-rate: In general, the charge/discharge currents of a cell are represented by C-rates in order to normalize against the capacity, which is usually quite different between cells. C-rate, by definition, is a measure of the rate at which a cell is discharged with respect to its full capacity. The meaning of 1 C is that a fully charged cell rated at 1 ampere hour (Ah) should constantly supply 1 A current for 1 hour. If the same fully charged cell is discharging at 0.5 C (C/2), it should deliver 500 mA current for 2 hours, and at 2 C, it should supply a current of 2 A for 30 minutes. The electrochemical performances of a cell at a rated current are generally measured using different battery analyzers. For instance, when a cell is subjected to different rated discharge using a battery analyzer, in practice a higher C-rate delivers a lower capacity and a lower C-rate produces higher capacity. These differences in energy with different C-rates occurred because of the internal losses, which in turn convert some of the energy into heat and thus reduce the output capacity at higher C-rates.

An E-rate, on the other hand, demonstrates the discharge power. For instance, a 1 E-rate is the required discharge power to discharge the whole cell in 1 hour. For a cell, normally, charging could be made by providing a constant current despite the variation in the voltage or by providing a constant voltage despite of the current drawn.

2.2.3.2 Variables to Describe the Battery Condition

State of Charge (SOC) (%): SOC is generally expressed in percentage, and it demonstrates the capacity of a battery at present as a percentage of the maximum capacity. In other words, it is a measure to demonstrate the capacity fullness of a battery. For instance, if the SOC of a battery is 100%, then it means that the battery is fully charged. Sometimes, the SOC is used to express the energy reserve of a battery.

Depth of Discharge (DOD) (%): Like SOC, DOD is also expressed in percentage, where it demonstrates the capacity of a battery that has been discharged and is expressed as a percentage of maximum capacity. In other words, it is a measure to express how deeply the battery is discharged. For example, if a battery is fully charged and its SOD is 100%, this means that the DOD of the same battery is 0%. Suppose the battery has used its 20% energy, and the 80% energy is reserved, then the DOD of the battery is 20%. Likewise, if a battery is 100% discharged then its DOD is 100%. The discharge of a battery to at least 80% DOD is normally referred to as a deep discharge. It is always recommended to keep the battery above this 80% DOD, otherwise it could lead to reduction in cycle life of the battery.

In general, SOC is used when describing the present state of a battery in use, while DOD is frequently used when describing the battery life after multiple uses. The formula connecting the SOC and DOD is shown below.

$$SOC = 100\% - DOD. \tag{2.1}$$

Terminal Voltage (V): It is the voltage difference across the two terminals of a battery or a load when the load is attached to the battery. In general, the terminal voltage of a battery is less than its electromotive force (EMF) value; this is because of the presence of the internal resistance of a battery, which in turn leads to the utilization of some of the EMF to overcome the resistance. The terminal voltage depends on three prime parameters, such as EMF, internal resistance, and load resistance. Thus, the terminal voltage can be expressed as follows.

$$V = EMF - IR, \tag{2.2}$$

where I represents the total current drawn out of the battery, and R is the internal resistance of the battery. The terminal voltage changes according to SOC and discharge/charge current.

Open-Circuit Voltage (V_{oc} or OCV): It is the difference in electrical potential among two terminals (anode and cathode) of the battery without any external current flow. This voltage developed as a result of electrochemical reaction differing according to the electrolyte and materials used to fabricate the cell. In a battery, the OCV is also known as the EMF, where it is the maximum voltage difference between the terminals when there is no external current and the circuit is open. For a battery, the voltage is also determined by the compatibility of the principle components of the cell, such as anode, cathode, and electrolyte. Thus, the OCV can also be defined as the difference in chemical potential between the anode (μA) and

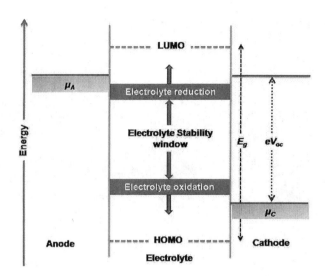

FIGURE 2.2
Schematic illustration of electrochemical potential window of sodium ion batteries.

the cathode (μC) (and is limited by the electrochemical window) (Figure 2.2). The OCV can be expressed as follows (equation 2.3). Where, e is the electronic charge.

$$V_{oc} = \frac{\mu A - \mu C}{e}. \tag{2.3}$$

Trickle Charge: Trickle charge is a low-level charging method, where a fully charged battery maintains charging under no external load (if the current is being drawn at the load, this charging process cannot maintain the battery at its fully charged state) at the rate equal to its self-discharge. This charging process makes sure that the battery will constantly be in its fully charged state. In general, this charging process is not advisable for organic electrolyte based LIBs and SIBs, mainly due to the safety issue associated with the overheating and possible explosion.

Internal Resistance: A practical battery is modeled as a linear circuit, where a voltage source is connected in series with a resistance. This resistance is called an internal resistance of the battery (equation 2.2). Normally, the internal resistance of a battery is dependent on multiple factors, such as the battery size, age, temperature, chemical properties, discharging current, etc. For instance, when the internal resistance increases, efficiency of the battery reduces, and the thermal stability gets decreased, because most of the charging energy is converted into heat. In general, it has two components, namely, electronic component because of the resistivity of the materials used to fabricate the battery and ionic component which is associated with the electrochemical factors like ionic mobility, conductivity of the electrolyte, active surface area of the electrode, etc.

Internal Resistance (IR) drop: The IR drop is the voltage drop that occurs between the end of the charge and the starting of the discharge due to the presence of internal resistance.

Overpotential: It is the difference between the thermodynamically calculated (theoretical) voltage and the experimentally obtained voltage under battery operating conditions. This is directly associated with the battery's voltage efficiency. The overpotential of a cell could adversely affect the important battery parameters, such as capacity and voltage. It is also noted that the overpotential of a battery during charge (deintercalation) and discharge (intercalation) process is dependent on SOC and DOD, respectively.

Voltage Hysteresis: Voltage hysteresis (voltage polarization) is the phenomenon observed between the charge and discharge curves of a battery. For a practical battery system, the charge voltage is considered to be greater than the discharge voltages because of the polarization emerging from the internal resistance of the battery materials. This hysteresis can be observed because of multiple reasons, such as the IR drop, electrochemical phase transition due to overpotential, structural changes of the material during charge/discharge cycles, etc.

Electrochemical Potential (voltage) Window: It is the safe voltage window that a battery can be operated in without compromising its performance and safety. In this potential range, the electrolyte molecule will not undergo oxidation or reduction processes. In general, it can be obtained by taking the difference between the reduction and oxidation potential. Figure 2.2 represents the schematic illustration of a safe electrochemical window of a battery. As discussed in many literatures, the energy difference between the Highest occupied molecular orbital (HOMO) and Lowest unoccupied molecular orbital (LUMO) level of an electrolyte is not in fact a stable electrochemical window for a battery. The concept of HOMO and LUMO are derived from approximated electronic structure theory while investigating electronic properties of isolated molecules, and their energy levels do not specify any species participating in redox processes.[15] The redox potentials are directly correlated with the difference in Gibbs free energy of the reactants and the products. Even though, in some specific cases, the redox potentials indicate a strong connection with HOMO energies; but the redox potential of the solvent can be also greatly affected by the presence of electrolytes and other molecules, leading to an offset as high as 4 eV from the HOMO. For instance, the HOMO energy level of commercially available battery solvents, such as ethylene carbonate and dimethyl carbonate are −10.51 eV and −9.64 eV, whereas the obtained oxidation potentials for the same solvents are −7.87 eV and −7.07 eV on the energy scale.[16] Hence, using the HOMO energy level to specify the electrolyte stability would result in overestimated electrolyte stability values. Thus, instead of using HOMO and LUMO to demonstrate the safe electrochemical window, it is better to use the voltage of electrolyte reduction at the negative potential and the voltage of solvent oxidation at the positive potential (Figure 2.2).

Battery Management System: A BMS is an electronic unit that monitors the parameters, such as voltage, temperature, etc., of a battery system to ensure proper functioning of the battery without compromising the safety. It protects each cell from overvoltage and overheating by shutting down the charging process. During the discharging, the BMS will help to disconnect the load when the cell is dropping below the minimum voltage level. In short, the BMS protects the battery from unsafe operating conditions.

2.2.3.3 Battery Technical Specifications

Nominal Voltage: It is the reported or reference voltage of a battery. This can be defined as the mid-point voltage or average voltage of a cell, which is generally measured when the battery is discharged to 50% of its maximum energy.

Cut-off Voltage: It is the minimum acceptable voltage of a battery. Normally, it is the voltage that describes the "empty" state (fully discharged) of a battery. When the battery is discharged to its cut-off voltage, the BMS helps the battery to stop its entire functioning. In general, the cut-off voltage of a battery is decided by the manufacturer, this is to ensure that the consumers can utilize the maximum capacity of the battery without compromising the subsequent battery performances or life.

Capacity or Nominal Capacity: It is the maximum stored charge, which can be drawn out of a fully charged battery under some particular discharge conditions. In other words, it is the maximum accessible ampere-hours when the battery is discharged from 100% SOC to its cut-off voltage under some specified discharge current or rate. It is generally obtained by multiplying the discharge current (A) with the discharge time (h) and has shown to decrease with increasing C-rate (current drawn). The specific capacity can be represented by both gravimetric specific capacity (Ah kg^{-1}) and volumetric specific capacity (Ah L^{-1}).

Cycle Life: It is the number of cycles that a battery can be charged and discharged within the given voltage range prior to the initial capacity falling to the specific stage (normally 80%) of the rated initial capacity. This cycle life is generally affected by the current (C-rate), DOD (higher DOD leads to poor cycle life), and atmospheric conditions such as humidity and temperature.

Irreversible Capacity: It is the capacity difference between the initial and the nth charge-discharge cycle. For a battery to achieve better available capacity, it is necessary to reduce its irreversible capacity loss. The major irreversible capacity loss occurred at the anode side during the initial few charge/discharge cycles, mainly due to the formation of a surface film (solid-electrolyte-interphase layer) at the interface between the anode and the electrolyte. This film is crucial for the proper functioning of a battery; however, the formation of solid-electrolyte-interphase layer normally consumes a large part of the potential capacity of the SIBs. Other reasons for the irreversible capacity loss is the irreversible secondary reactions happening at the electrodes due to multiple reasons, which include structural/phase changes of the active material with prolonged cycling. Also, it is revealed that the permanent lodging of the shuttling sodium ions at the anode side and resulting concentration decrease of the sodium ion at the cathode side cause the irreversibility and hence the reduction in capacity over multiple usage of the cell.

Self-Discharge: This is a restorable capacity loss, due to the internal chemical reactions happening in the battery without any external connection (however, comparable to the application of a small external load) between the anode and cathode terminals, even when the cell simply sits on the shelf (thus causes decrease in shelf-life of the batteries). In general, it describes the percentage of rated capacity diminished per month at a specified temperature. In detail, this phenomenon is just like a normal chemical reaction (closed circuit discharge) in batteries. It is believed that for batteries like LIBs and SIBs, the rate of self-discharge will get reduced over time due to the formation of a passivation layer at the electrodes after some time. At elevated temperatures, the rate of self-discharge will be more, thus it is advisable to store the battery at lower temperatures. Apart from the ambient temperature, the rate/fastness of the self-discharge is dependent on multiple factors, such as, battery type/chemistry, SOC, charging current/mode, etc. In general, the rechargeable (secondary) batteries have higher self-discharge rate than the non-rechargeable (primary) batteries.

Coulombic Efficiency: Coulombic efficiency is also known as faradaic/current efficiency, defines the ratio of the total charge that exits the battery during the discharge to the total charge that enters into the battery during the charging process. All batteries have Coulombic efficiency losses. There are multiple reasons for this loss, which include parasitic/secondary reaction that happens within the battery, ultra-fast/high-rated charging, heavy loading, etc. It is possible to attain high Coulombic efficiency for rechargeable batteries like LIBs and SIBs. However, this can be achieved only by charging the battery at moderate rate/current at low temperature. For instance, high rates of charging cause lowering of the Coulombic efficiency due to the losses associated with the charge acceptance and heat, also at a very low charge rate, the self-discharge comes into picture and thus reduces the Coulombic efficiency.

Energy or Nominal Energy (Wh): It is the total watt-hours of a battery when it is discharged to the cut-off voltage from the 100% SOC under a specific discharge current (C-rate). The energy is estimated by multiplying the discharge power (W) with the total discharge time (h). The energy also shows the same trend as capacity with the C-rate, that is the energy decreases with the increasing C-rate. It can be expressed in two ways: (1) volumetric energy expressed in Wh L^{-1}, generally termed as energy density and (2) gravimetric energy (Wh kg^{-1}) also called specific energy.

Specific Energy (Wh kg^{-1}): Specific energy/gravimetric energy density is the nominal battery energy per unit mass, sometimes referred to as the gravimetric energy density. It is the characteristic of a battery's electrochemistry and packaging.

Energy Density (Wh L^{-1}): Energy density/volumetric energy density is the nominal battery energy per unit volume in liters. It describes the required size of a battery to achieve a targeted performance.

Power Density: The power performance of a battery can be defined in terms of power density/volumetric power density [Watts per liter (W L^{-1})] and gravimetric/specific power density [Watts per kilogram (W kg^{-1})].

Specific Power (W kg^{-1}): It is the maximum accessible power per unit mass. Like specific energy, the specific power is also defined as the characteristic of the cell/battery's electrochemistry and packaging. That is, it decides the weight of a cell/battery which is required to achieve a targeted performance.

Power Density (W L^{-1}): It is the maximum accessible power per unit volume and describes the required size of a battery to achieve a targeted performance.

2.3 Electrochemical Reaction Mechanism

In the alkali metal series, sodium is the second lightest and smallest material (as lithium is the lightest and smallest), and since the electrochemical storage mechanism of the SIBs are similar to LIBs,[17] the energy density sacrifice is considered to be minimal. Also, the cost can be reduced by substituting aluminum substrate at the anode side instead of copper, because unlike lithium, sodium metal does not form an alloy with aluminum. Even though SIBs bring some cost reduction, the 3.3 times higher atomic mass and 0.3 V lower standard reduction potential of sodium as compared to the lithium counterpart is always an issue when it comes to the commercial aspect. However, as compared to other emerging batteries, SIBs have the second highest operating potential of over 3 V. Even though the atomic mass of sodium is 3.3 times more than that of lithium, the formula mass of a similar material, for e.g., $NaCoO_2$ (114 g mol^{-1}) is only 16% heavier than the lithium counterpart [$LiCoO_2$ (98 g mol^{-1})]. Thus by considering a cell with C_6//$LiCoO_2$ and C_6//$NaCoO_2$ configuration, the net gain in weight due to sodium substitution is only ~9% as compared to lithium, leading to a very small difference in specific capacities as compared to LIBs.[17]

Figure 2.3 illustrated the different components of SIBs that are responsible for the charge/discharge phenomenon and hence the storage. The main redox process in SIB is based on the sodium ion intercalation/de-intercalation among the anode (negative electrode) and cathode (positive electrode). Both these electrode materials are coated on the appropriate substrate/current collector. In general, the cathode material is coated on the aluminum foil and the anode material is applied on the copper substrate; however, for SIBs, the anode material can also be coated on the aluminum substrate, thus can reduce the cost of electrode production. The electrodes are ionically connected by

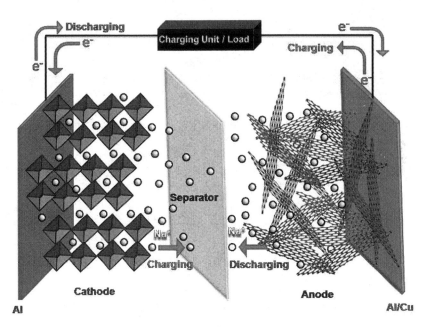

FIGURE 2.3
Schematic representation of electrochemical working process of sodium ion batteries.

the porous membrane (separator) immersed in the electrolyte which is placed between the electrodes; this is to ensure appropriate ionic transportation and at the same time to block any electrical contact (short circuiting) between the anode and cathode. The potential anode, cathode, and electrolyte for SIBs are discussed in detail in the subsequent chapters. During the charging process, the sodium atoms at the cathode side release the electrons to the external circuit and thus turn into sodium ions, then the formed ions move toward the anode through the electrolyte and combine with the electrons from the external circuit at the anode side. As a result of this process, the electrical energy provided by the external power source (during the charging) gets converted and stored in the form of chemical energy in the battery. During the discharge process, the abovementioned process gets reversed. That is when the battery is discharged through an external load, the sodium ions de-intercalate from the anode and travel through the electrolyte and reach the cathode side, while at the same time, the electrons move toward the external load from the anode side.[18] To convert and store energy in SIBs through the abovementioned electrochemical processes; the SIB's components must have some basic requirements. For instance, the cathode material should have appropriate lattice sites or spaces to lodge and release the sodium ions freely in a reversible manner. Also, the cathode should have stable crystal structure and adequate number of active sites to ensure good cycle life and capacity, respectively. The cathode having high operating potential is also a very favorable attribute to have better energy performances. On the other hand, for the anode, it is beneficial to have a lower voltage, thus, to ensure the high voltage and energy when it is assembled in full cell. Like the cathode, the anode material should also have enough numbers of void spaces to house sufficient sodium ions in order to achieve high specific capacity. The material should also possess a stable structure to tolerate the volume expansion during the reversible sodium ion intercalation/de-intercalation mechanism. Apart from the abovementioned properties, both the electrode materials should have chemical stability under high power application and over a range of operational temperatures. Thus, with the abovementioned properties and appropriate electrolytes and separator, the assembled battery could effectively convert and store the chemical energy into electrical energy.

References

1. Liu, C., Neale, Z. G., Cao, G. Understanding electrochemical potentials of cathode materials in rechargeable batteries. *Mater. Today* **2016**, *19* (2), 109.
2. Agency, I. E. *Transport Energy and CO_2: Moving towards Sustainability*, **2009**, OECD, Paris, France.

3. Wang, C., Xu, Y., Fang, Y., Zhou, M., Liang, L., Singh, S., Zhao, H., Schober, A., Lei, Y. Extended π-conjugated system for fast-charge and -discharge sodium-ion batteries. *J. Am. Chem. Soc.* **2015**, *137* (8), 3124.
4. Zhang, Q., Uchaker, E., Candelaria, S. L., Cao, G. Nanomaterials for energy conversion and storage. *Chem. Soc. Rev.* **2013**, *42* (7), 3127.
5. Kalyanasundaram, K., Grätzel, M. Themed issue: Nanomaterials for energy conversion and storage. *J. Mater. Chem.* **2012**, *22* (46), 24190.
6. Singh, R., Setiawan, A. D. Biomass energy policies and strategies: Harvesting potential in India and Indonesia. *Renew. Sust. Energ. Rev.* **2013**, *22*, 332.
7. Tarascon, J. M., Armand, M. Issues and challenges facing rechargeable lithium batteries. *Nature* **2001**, *414*, 359.
8. Wang, Y., Liu, B., Li, Q., Cartmell, S., Ferrara, S., Deng, Z. D., Xiao, J. Lithium and lithium ion batteries for applications in microelectronic devices: A review. *J. Power Sources* **2015**, *286*, 330.
9. Van Noorden, R. The rechargeable revolution: A better battery. *Nature* **2014**, *507* (7490), 26–28.
10. Armand, M., Tarascon, J. M. Building better batteries. *Nature* **2008**, *451*, 652.
11. Lu, L., Han, X., Li, J., Hua, J., Ouyang, M. A review on the key issues for lithium-ion battery management in electric vehicles. *J. Power Sources* **2013**, *226*, 272.
12. Choi, J. W., Aurbach, D. Promise and reality of post-lithium-ion batteries with high energy densities. *Nat. Rev. Mater.* **2016**, *1*, 16013.
13. Duduta, M., Ho, B., Wood, V. C., Limthongkul, P., Brunini, V. E., Carter, W. C., Chiang, Y.-M. Semi-solid lithium rechargeable flow battery. *Adv. Energ. Mater.* **2011**, *1* (4), 511.
14. Howard, W. F., Spotnitz, R. M. Theoretical evaluation of high-energy lithium metal phosphate cathode materials in Li-ion batteries. *J. Power Sources* **2007**, *165* (2), 887.
15. Peljo, P., Girault, H. H. Electrochemical potential window of battery electrolytes: The HOMO–LUMO misconception. *Energ. Environ. Sci.* **2018**. doi:10.1039/C8EE01286E 10.1039/C8EE01286E.
16. Barnes, T. A., Kaminski, J. W., Borodin, O., Miller, T. F. Ab Initio characterization of the electrochemical stability and solvation properties of condensed-phase ethylene carbonate and dimethyl carbonate mixtures. *J. Phys. Chem. C* **2015**, *119* (8), 3865.
17. Kubota, K., Komaba, S. Review—Practical issues and future perspective for Na-ion batteries. *J. Electrochem. Soc.* **2015**, *162* (14), A2538.
18. Pan, H., Hu, Y.-S., Chen, L. Room-temperature stationary sodium-ion batteries for large-scale electric energy storage. *Energ. Environ. Sci.* **2013**, *6* (8), 2338.

3

Anode Materials for Sodium Ion Batteries

3.1 Introduction

Different from the cathode materials in Sodium ion battery (SIBs), the exploration for appropriate anode candidates in SIBs is more difficult and complex, because very few materials have been reported to be useful as negative electrodes. For example, most of the previous reported cathode materials for Lithium ion battery (LIBs) can be used as cathode materials in SIBs just by replacing the lithium by sodium. However, as for the anode materials in SIBs, most of the current used anodes, such as the graphite which is a traditional commercial anode material in LIBs, cannot be used as the anode in SIBs directly because the unstable formation of the solid electrolyte interphase (SEI). Therefore, the exploration for novel commercial anodes in SIBs is very important.

As shown in Figure 1.2, the present researched improvements on anodes in SIBs have been obtained by employing the metal oxides, metal alloys, and carbonaceous materials as anodes. In addition, according to the reaction mechanism differences during Na^+ insertion/extraction processes among different candidates, the abovementioned anodes can be divided into three different groups: (1) the insertion reaction, (2) the conversion reaction, and (3) the alloying reaction. In the following sections, detailed discussion about these three different type anodes will be provided.

3.2 Insertion Materials

For the insertion materials, the research attention mainly focuses on two different anode materials, the carbon- and titanium-based materials. Compared with other anode materials, the main advantage of carbon/titanium-based materials is their unique structure ability which is favorable for Na^+ ions intercalation. For example, a reversible capacity about 300 mA h g^{-1} can be obtained for the hard carbon anodes. Apart from that, the low-cost and low operating potential (~0 V vs. Na^+/Na) are also their selling points which made them promising as ideal anode candidates for SIBs.

However, their drawbacks such as poor reversibility and reversible capacity should not be neglected because the current electrochemical performance of carbon/titanium-based materials is still far from that achieved by graphite in LIBs. More importantly, the limited understanding about its Na+ storage mechanism in a disordered carbon structure is another big challenge for carbon-based anode development in SIBs.

3.2.1 Carbon-Based Anode Materials

3.2.1.1 *Graphite*

Graphite microspheres (GMs), graphite microsized particles, and mesophase carbon microbeads (MCMBs) with a high graphitization degree are dominating the commercial anode market in LIBs. The Li+ storage mechanisms in this kind of ordered carbon structure were well established. Li+ can be inserted among the adjacent graphene layers during the electrochemical reduction process. As a result, the Li ion-inserted graphite compounds are generated via the stage transformations in the Li-based Energy Storage System (EES).[1-2] Due to the successful application of graphite in the LIBs, it is interesting and important to explore the electrochemical performance between graphitic carbon and Na ions. This work was first conducted in the 1980s based on the graphite/polyethylene oxide $NaCF_3SO_3$/Na cell (Figure 3.1a)[3] and confirmed that the sodiation behavior of graphite is difficult to obtain and the electrolyte exhibited a serious degradation (Figure 3.1b).[4] The first principles calculation data of the formation energy for Na-graphite showed that it is difficult to proceed with the sodiation reaction inside graphite due to the energetic instability of the Na-graphite. The distance between the adjacent graphene layer will be reduced during the Na ions insertion process which is mainly due to the thermodynamic instability of the newly formed Na-graphite compound. The ionic radius of Na+ is ~0.3 Å larger than that of the Li+, which means that it is more difficult to conduct the insertion of Na+ into the graphite interlayers, leading to the unstable structure and poor electrochemical activity.[5] For example, during the first stage of the Na+ intercalation into graphite within SIBs, the thermodynamically unstable NaC_6 and NaC_8 will be produced, resulting in the limited specific capacity of graphite.[5,6]

Although it is difficult to realize high and stable reversible capacity for the graphite anode in SIBs, a breakthrough on the intercalation between solvated Na+ ions and graphite was successfully obtained during the past few years. For example, a ternary Na-graphite compound was formed by Jache and Adelhelm using the solvated-Na-ion intercalation.[6] This can be explained by the following equation: $C_n + e^- + A^+ + y\,\text{sol}\,v \leftrightarrow A^+(\text{sol}\,v)_y C_n^-$.[6] Zhu's group conducted similar research and confirmed that a stage evolution process was proceeded during the solvated Na+ insertion into graphite, resulting in the ternary graphite intercalation compounds.[7] The electrolyte also plays a very important role during the sodiation process. For example, ether-based electrolytes which are endowed with a great number for donation are able to

Anode Materials for Sodium Ion Batteries

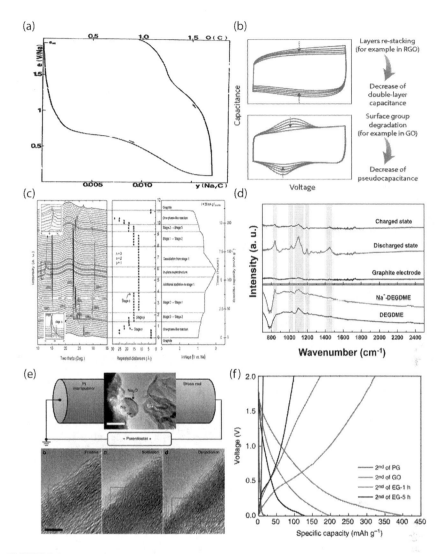

FIGURE 3.1
(a) Variation of graphite electrode potential during reduction and oxidation at constant current, electrolyte: (PEO-NaCF$_3$SO$_3$), discharge rate: (C/200, temperature: 82°C) (Copyright 1988, with permission from Elsevier). (b) Generic voltammetric behavior of graphene-based electrochemical capacitors over prolonged cycling. Top: Effect of graphene layers re-stacking (such as in rGO) on the double-layer capacitance. Bottom: Effect of surface group degradation (such as in GO) on pseudocapacitance (Nature Publishing Group, Copyright 2015). (c) In operando synchrotron X-ray diffraction analysis of the structural evolution of the ternary Na-ether-graphite system (Copyright 2015 The Royal Society of Chemistry). (d) FTIR analysis demonstrates that Na storage occurs through solvated Na ion co-intercalation (Copyright 2015 Wiley-VCH Verlag GmbH & Co. KGaA). (e) In situ TEM investigation of the sodium storage mechanism in the EG-1h sample. (f) Electrochemical performances of PG, GO, EG-1h, and EG-5h in organic 1.0 M NaClO$_4$ in a polycarbonate solvent liquid electrolyte (Nature Publishing Group, Copyright 2014).

produce the stable and no-polar Na ions solvated compounds, leading to the easy way to react with graphite.[8–10] With the help of *in-situ* X-ray diffraction, K. Kang's group explored the electrochemical behavior between graphite and solvated Na$^+$, and they found that there is a phase transition which can be reversible during the following charge/discharge process (Figure 3.1c).[10] Kang's group studied the differences of the Na ions storage performance between the solvated Na$^+$ and graphite under different electrolytes and Na-based salts, such as ethylene carbonate/diethyl carbonate (EC/DEC), ether-based electrolytes, NaClO$_4$, and NaCF$_3$SO$_3$. During the Na ions insertion process, the electrolyte decomposition can be effectively suppressed by the ether-based electrolytes, leading to very thin SEI films produced on the outer surface of the graphite which benefits the Na-ion-solvent transport into the interlayer of the graphite lattice. On the other hand, the carbonate-involved electrolytes, such as EC/DEC, prefer to generate thicker SEI layers on the graphite surface because of the decomposition of the electrolytes during the Na ions insertion stage, which suppresses the Na$^+$ transportation inside the graphite (Figure 3.1d). Therefore, a reversible capacity of 150 mAh g^{-1} and outstanding cycling stability for 2500 cycles were obtained for the graphite anode in SIBs, using the ether-based electrolyte which is dissolved with NaPF$_6$ salt. Apart from the effects derived from electrolytes, the natural characteristics of graphite are also crucial for its sodiation behavior. Wang's group explored the electrochemical performance of the expanded graphite (EG) as an anode in SIBs.[11] They found that the graphite with a larger interlayer lattice distance contributed higher reversible capacity (284 mAh g^{-1}) and long cycling life over 2000 cycles. According to the *in-situ* TEM results, reversible Na ion-insertion/extraction happened during the discharge/charge process, respectively (Figure 3.1e and f). In addition, Kang's group also studied the Na ions intercalation performance of the expanded graphite oxide (GO) in SIBs. They reported that the functional groups on the GO have a strong relationship with its electrochemical abilities.

3.2.1.2 Hard Carbon

Compared with graphite-based carbon, the sodium storage ability of the hard carbon is more promising. The electrochemical performance of Na storage performance inside the non-graphitic carbon was first reported by Doeff's group in 1993. They explored the Na ions sodiation/desodiation reactions inside the hard carbon which was derived from the pyrolysis of petroleum coke (Figure 3.2a).[12] According to their research, Na$^+$ had the ability to conduct the insertion/extraction reaction into the disordered carbon. Therefore, it is possible to employ hard carbon as the anode material in SIBs. In addition, the electrochemical reaction mechanism was also studied by this group (Figure 3.2b).[13] It was found that the insertion of Na ions was mainly conducted in two positions. Compared with the graphite with smaller sized distance between the adjacent graphene sheets, it is

Anode Materials for Sodium Ion Batteries

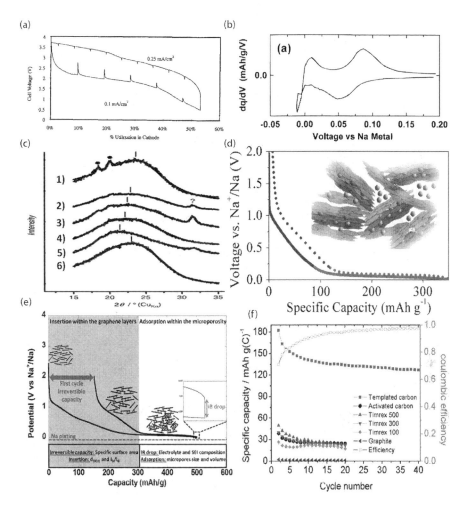

FIGURE 3.2
(a) First cycle of the cell, 0.1 mA cm^{-2} discharge, 0.25 mA cm^{-2} charge, temperature = 100°C (Copyright 1993 The Electrochemical Society). (b) Differential capacity versus voltage for Na/carbon cells using glucose pyrolyzed to 1150°C (Copyright 2000 The Electrochemical Society). (c) Ex situ XRD patterns of hard-carbon electrodes: (1) pristine electrode, galvanostatically reduced to (2) 0.40 V, (3) 0.20 V, (4) 0.10 V, (5) 0.00 V, and (6) oxidized to 2.00 V after reduction to 0.00 V polyvinylidene difluoride [PVdF] binder) (Copyright 2011 Wiley-VCH Verlag GmbH & Co. KGaA). (d) Visual representation of the card-house model on Na-ion storage in hard carbon. The two distinct phases: intercalation inside turbostratic nanodomains (TNs) and pore filling are seen (Copyright 2015 American Chemical Society). (e) Typical potential vs. capacity profile of hard carbon when tested against sodium metal counter electrodes (Copyright 2015 The Electrochemical Society). (f) Specific capacities (C/5, Na-insertion) upon cycling for different carbons and Coulombic efficiency for the templated carbon. All data are obtained at room temperature at C/5 (Copyright 2012 The Royal Society of Chemistry).

easier for Na ions to insert inside the hard carbon due to its increased lattice plane distance. Furthermore, the disordered carbon is originally endowed with more nanosized void spaces inside and also an increased amount of nanopores were generated during the heat treatment of soft carbon, resulting in increased positions for Na ions insertion.[14] Based on their research, a very high reversible capacity (300 mAh g^{-1}) and low operating voltage platform (around 0 V) were both obtained using glucose-based hard carbon as anode in half SIBs. After the carbonization of glucose under inert gas atmosphere, the hard carbon with rich nanopores and disordered graphene layers inside was produced, benefiting its high specific capacity. Doeff's research illuminated the way of application of hard carbon-based as anode materials in SIBs, and that inspired the following research in this area. In 2011, Fujiwara's group systematically explored the relevant structural changes of hard carbon during the Na ions insertion.[15] They found that the distance between the adjacent graphene sheets was significantly increased if more Na ions continuously inserted into the hard carbon below 0.2 V, resulting in the position of XRD peaks moved to lower angles (Figure 3.2c). Moreover, the direct evidence from the small angle XRD demonstrated the reversible insertion/extraction of Na ions in the voltage range of 0–0.2 V within the inside nanopores.

According to the above research work, it can be found that the enlarged interspace between parallel graphene layers contributed to the large reversible Na ions insertion/extraction process, leading to the high reversible capacity for a hard-based anode in SIBs. For example, 0.446 Å of the interlayer distance for the hard carbon (*d spacing*: 3.354 Å) was increased when compared with the graphite (*d spacing*: 3.8 Å), and this enlarged interspace provided enough room for the insertion of Na ions. Therefore, according to the Na NMR study, there are two main reversible reactions within the carbon anode-based half SIBs. The first one is the reversible insertion/extraction of Na ions with the disordered graphene sheets and another one is the reversible reaction with the inside nanopores.

The first report about the high capacity carbon-based SIBs anode was proposed in 2011 by Adelhelm's group.[16] They found that a high reversible capacity around 100 mAh g^{-1} at C/5 was obtained. The pitch carbon was employed in their work as the carbon precursor and created the hierarchal porosity structure inside after using the nanocasting route. As a result, the disordered graphene sheets and the nanosized pores were combined together in one system, which contributed the large reversible capacity during Na ions insertion/extraction process. In addition, they also studied the electrochemical differences among hard carbon materials with different surface areas and pore volumes, and they found that the hard carbon with the highest surface area and pore volume showed the lowest reversible capacity. Therefore, the high capacity for a hard carbon-based anode in SIBs cannot be obtained by simply increasing the surface area or by enhancing the degree of porosity. A suitable porous structural design is crucial for the electrochemical performance of a hard carbon anode in SIBs. Interestingly, another research group further confirmed that the high reversible capacity is not only effected by the enlarged interlayer distance among the graphene sheets, but also has a tight relationship with the vacancy defects inside the hard carbon. Both of the large interlayer distance and vacancy defects contribute together for the active reversible Na ions intercalation. Due to the existence of the vacancy defects, a strong ionic binding was generated between the defects and Na ions, leading to the enhanced Na ions intercalation within the hard carbon, benefiting the rate performance and cycling stability.[17] Moreover, Ji's group also explored the Na ions storage ability at these vacancy defects and proposed a new Na ions storage mechanism.[18] Based on their study, the Na ions can also be intercalated within the hard carbon at the low voltage platform area, which demonstrated that there was a three-step reaction process rather than two-step reaction process (Figure 3.2d–f). In conclusion, the Na ions intercalation mechanism can be divided into three different stages: the Na ions absorbed by the vacancy defects under the low voltage area (the sloping region in the charge-discharge curve), the Na ions insertion/extraction with the graphene sheets, and the Na ions absorbed by the nanosized pores inside the hard carbon-based anode materials in half SIBs. Unfortunately, limited research work and evidence have been reported on the Na ions storage mechanism till now, and further experimental study must be launched to illuminate the relevant reaction mechanism of the hard carbon materials.

3.2.1.3 Graphene

Graphene, as a novel two-dimensional (2D) carbon material, has attracted tremendous attention in the past few years due to its unique advantages, for example, the remarkable electrical, chemical, and mechanical properties.[19] The abovementioned advantages provide fast ion diffusion within the shortened transfer channels and contribute large fresh edges giving increased ion extraction/insertion pathways.[20] As a result, graphene is regarded as a promising anode material in energy storage systems, such as the LIBs and SIBs.[4,21] In addition, graphene is also a promising matrix for different materials to contribute outstanding electrochemical performance, such as metal oxides, silicon, sulfur, etc. Taking advantages of graphene, the Li or Na ions storage ability is significantly improved. The reversible Na ions insertion/extraction can be conducted in the reduced graphene oxide (rGO). Many groups have published relevant reports and proposed reasonable reaction mechanisms during these processes.[22-26] According to a report from Wang's group, rGO is more promising to be used as an anode in SIBs because it has an improved electronic conductivity and increased number of active positions which contribute larger reversible capacity during the sodiation/desodiation process. Moreover, compared with graphite, rGO is endowed with a larger interlayer distance which helps the Na ions with an accelerated insertion/extraction. As a result, a reversible capacity of about 140 mAh g^{-1} for over 1000 cycles was obtained for the rGO anode.[22] In addition, the graphitic degree for the graphene material is also crucial for the Na ions storage ability. The Na ions storage performance of various graphene materials with different graphitic degrees was studied by Ding.[27] They chose different carbonization temperatures (600°C–1400°C) during the heat treatment stages and obtained graphene materials with relevant different graphitic degrees. They found that significant improvement of the Na ions storage ability could be obtained when the heat temperature was over 1100°C, which was mainly due to the increased interlayer distance (0.388 nm). Apart from the interlayer distance, the defect amount also plays an important role for the Na ions insertion, because the Na ions could be easily absorbed by the defects, leading to extra active positions and higher specific capacity.

3.2.1.4 Heteroatom Doping

Heteroatoms, such as nitrogen (N), sulfur (S), phosphor (P), and boron (B), can significantly improve the electrochemical performance of the carbon-based anode materials. The main reason for the performance enhancement is derived from the increasing number of the defect position and functional groups inside the carbon-based materials after the hetero atom doping treatment. The increasing quantity of the defects promotes the absorption of Na ions, leading to the high specific capacity. In addition, the introduced new functional groups on the surface of the carbon materials improve the activity reaction between electrode-electrolyte.

In detail, the N doping contributes to the Na ions transfer process when the N atoms are doped inside the carbon materials.[28,29] A self-support elastic carbon-based film which was composed of the N-doped porous carbon nanofibers was created by Wang.[30] This flexible carbon film showed outstanding electrochemical performance with a high 212 mAh g^{-1} reversible capacity at 5 A g^{-1} current density. Moreover, it also exhibited a good cycling stability with a long cycling life over 7000 cycles (Figure 3.3a). Just recently, S-doped amorphous carbon material was reported as the anode in SIBs, exhibiting very high capacity and excellent rate ability. The high capacity is derived from the additional active reaction positions after the S doping, which provides extra sites to absorb Na ions. The improved rate performance is due to the enlarged interlayer distance which enables the Na ions insertion/extraction with the carbon structure more easily. Another S-doped disordered carbon composite with high sulfur content was also proposed by another research group. This novel composite is endowed with a unique 3D coral-like morphology. Because of the high S percentage inside this composite, as high as 26.9%, a high specific capacity of 516 mAh g^{-1} was obtained. However, a relatively poorer cycling stability was hardly avoided for this material if compared with the previous mentioned S-doped carbon material, due to the higher sulfur quantity inside (Figure 3.3b).[31]

Anode Materials for Sodium Ion Batteries 31

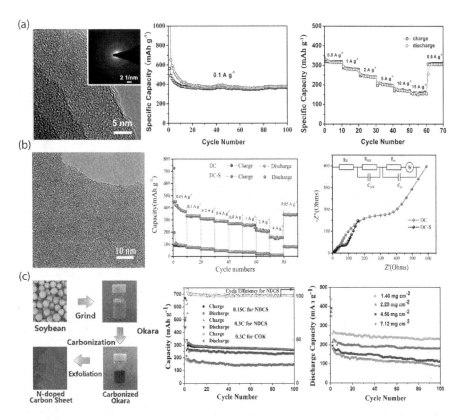

FIGURE 3.3
(a) High-resolution transmission electron microscopy (HRTEM) of N-CNF (inset is the selected area electron diffraction [SAED]) (left). Electrochemical performance of N-CNF, cycling performance at a current density of 0.1 A g^{-1} (middle) and rate capability test (right) (Copyright 2015 Wiley-VCH Verlag GmbH & Co. KGaA). (b) HRTEM image of sulfur-doped disordered carbon (DC-S) (left). Electrochemical performances of DC-S (right) (Copyright 2015 The Royal Society of Chemistry). (c) The preparation process of N-doped disordered carbon (NDCS) derived from okara (left). Electrochemical performances of DC-S (right) (Copyright 2015 Wiley-VCH Verlag GmbH & Co. KGaA).

3.2.1.5 Biomass-Derived Carbon

According to all the above discussed carbon-based anode materials for SIBs, it can be seen that the most promising candidate anode material is the amorphous carbon materials with disordered graphene sheets inside. The disordered graphene sheets are endowed with larger interlayer distance and increased defects, both of which contribute to the high reversible capacity and enhanced rate ability.

Biomass-derived amorphous carbon materials have attracted increasing research attention during the past few years due to their low-cost, easy accessibility, and relative facile fabrication process.[27,32–37] Moreover, from the renewable and green point of view, employing biomass-derived carbon material is crucial for future energy development. For example, the carbon materials which are derived from wood have several advantages, such as binder-free, current collect-free, self-supporting, flexible, and low-cost.[32] Hu's group made great efforts on the wood-derived electrode materials during the past 5 years. A wood-derived carbon sheet with disordered structure inside was reported, and this carbon material showed very high mass loading (55 mg cm^{-2}) and reversible capacity (mAh cm^{-2}). Just recently, Yan employed the okara as the original carbon precursor to produce a unique high percentage N-doped carbon material.[36] After the carbonization treatment for okara, the inside carbon sheets were expanded, leading to the increased interlayer distance and high surface area. In addition, the original N content inside okara is high, which benefits the nitrogen content inside the final product. Taking advantage of the high N content and enlarged interlayer distance, this okara-derived disordered carbon material exhibited a high reversible capacity of about 293 mAh g^{-1} and a long cycling life of 2000 cycles (Figure 3.3c). Based on these reports, it is promising to employ the biomass-derived carbonaceous materials as an anode for SIBs.

Unfortunately, the initial Coulombic efficiency (CE) is relatively low for these disordered carbon materials due to the porous structure which increases the surface area. The larger surface area promotes the formation of extra SEI films, leading to increased capacity loss after the first cycling test. As a result, CE is crucial for the commercial full battery application, because it significantly limits the enhancement for the practical battery energy density. Apart from the low CE during the initial cycling test, the rate performance and top density of the biomass-derived carbon anodes are also poor due to their porous structure. In order to obtain a higher energy density and reversible specific capacity, relatively low surface area and total pore volume for the carbon-based anode materials are necessary for the practical application. Decreasing the capacity loss during the cycling is a "multiparty" work, which includes the use of suitable binders, electrolytes, and conductive agents. Furthermore, a reasonable design of the structure, morphology, and size distribution of the active materials is also crucial to improve the rate performance. From the commercial application point of view, the low-cost, facial fabrication process, high mass loading electrodes, environmental friendly ability, and extensive understanding about the real reaction mechanism during the Na ions insertion/extraction process are all required.

3.2.2 Titanium-Based Materials

Generally, sodium dendrite and the relevant metallic sodium plating will be easily generated on the surface of anode materials in SIBs due to the low voltage platform, resulting in serious safety issues during the practical application.[13,14,18,38] According to the relevant research history in LIBs, it is also promising to employ metal oxide materials as the anode in SIBs.

Among various metal oxides, titanium-based oxides attract tremendous research attention because of the suitable voltage platform, low-cost, and environmental friendly ability.[38,39] Currently, the research is mainly focused on three main types of the titanium-based oxides, such as the sodium-titanate compounds,[38,40–52] spinel-lithium titanate,[53–65] and titanium dioxides.[53,66–92] The invertible reaction during the charging/discharging process in SIBs is original from the $Ti^{4+/3+}$ redox couple. During the past 3 years, a great number of relevant studies were focusing on exploring the sodiation/desodiation mechanism along with enhancing their reversible capacity, rate ability, and initial CE.

3.2.2.1 TiO₂

TiO₂ can be divided into several types based on its crystal structure, such as rutile-TiO₂,[86–89,93] anatase-TiO₂,[85] bronze-TiO₂,[94] and brookite-TiO₂ (Figure 3.4a–h).[90] Compared with others, anatase TiO₂ can be the most promising candidate when it is used as an anode in SIBs, because it has the similar activation barrier to process sodiation as lithium.[301–303] The TiO₂ nanotubes were prepared for the first time by Xiong's group and showed a promising performance for SIBs (Figure 3.5a). However, the sodiation process cannot happen if the TiO₂ has a high crystalline or microsized diameter. Both the high crystal degree and large size distribution limit the Na ions insertion/extraction due to the larger ionic size of Na ions. For example, the high crystal degree means the material is endowed with a reduced interlayer distance

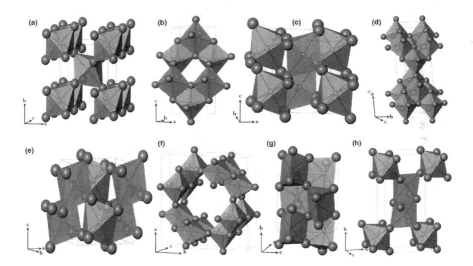

FIGURE 3.4
Crystal structures of TiO₂. (a) Rutile, (b) anatase, (c) bronze, (d) brookite, (e) columbite, (f) hollandite, (g) baddeleyite, and (h) ramsdellite phases (Copyright 2016 The Royal Society of Chemistry).

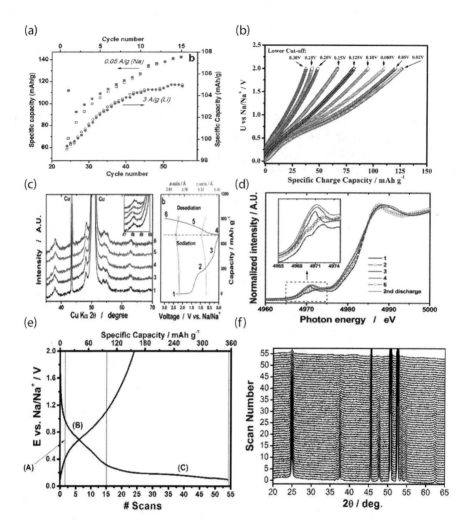

FIGURE 3.5
(a) Electrochemical characterization of TiO₂NT in Na half-cell. Specific capacities as a function of cycle number measured at a current density of 0.05 A/g in Na system: solid squares represent discharge and black open squares represent charge; and at 3 A/g in Li system: solid circles represent discharge and black open circles represent charge (Copyright 2011 American Chemical Society). (b) Galvanostatic investigation of the influence of the lower cut-off potential on the Coulombic efficiency for the first cycle: The upper cut-off potential is kept constant at 2.0 V, while the lower cut-off potential is set to 0.02, 0.05, 0.085, 0.10, 0.125, 0.15, 0.20, 0.25, and 0.30 V (Copyright 2014, with permission from Elsevier). (c) Ex situ XRD patterns and (d) X-ray Absorption Near-Edge Structure (XANES) K-edge spectra of 2.9 wt% carbon-coated anatase nanorod TiO₂ (Copyright 2014 American Chemical Society). (e) In situ XRD analysis of the (de-)sodiation mechanism of anatase TiO₂ nanoparticles and (f) XRD patterns obtained in situ upon discharge (sodiation) of the TiO₂-based electrode composite (Copyright 2015 Wiley-VCH Verlag GmbH & Co. KGaA).

between the parallel planes, which makes the Na ions diffusion more difficult. In addition, the larger sized distribution of the active composite can lead to the longer diffusion pathway for electrons and Na ions. Therefore, the promising approaches to achieve enhanced electrochemical performance are to reduce both the crystal degree and particle size. Usually, reducing the particle size from microsize to nanosize is the most important way to realize the abovementioned targets. Apart from the natural characteristics of the active materials, the electrolyte composition and working potential range are also crucial for the final performance (Figure 3.5b).[71] Gonzalez explored the SEI formation process of TiO_2 when the discharge voltage was over 0.3 V during the Na ions initial insertion stage.[73] The surface modification of anatase TiO_2 also has an important effect on the cycling stability and rate performance. Moreover, similar to the prelithiumation technique in LIBs, presodiation is also a promising way to improve the CE during the initial cycling processes. Therefore, employing the presodiation technique to the TiO_2 anode can effectively reduce the irreversible reaction and capacity losses.

According to Passerini's report, the reversible sodiation/desodiation of TiO_2 can only be obtained within the amorphous sodium titanate phase which is generated after the Na ions insertion during the discharging process. However, the reversible sodiation process cannot be conducted in other typed reaction processes, such as the sodium superoxide, metallic titanium, and oxygen evolution (Figure 3.5e and f).[74] Usui found that the reversible desodiation/sodiation can be performed in the crystal rutile-TiO_2[88] and a certain amount of Nb doping on rutile-TiO_2 can significantly enhance the electrical conductivity and rate performance. The TiO_2 microspheres which were composed of the rutile TiO_2 nanoneedle clusters showed an outstanding cycling stability during the 200 cycling tests.[89] In addition, the crookite-TiO_2 with a low crystalline degree which is available for the reversible Na ions insertion/extraction can also be used as the anode for the SIBs.[90]

Unfortunately, the sodium kinetics for these TiO_2 materials are still sluggish because of Na ions with a large ionic size. Reducing the particle size to nanosize and introducing carbon coating layers are effective ways to overcome this disadvantage. The nanosized particle has the short diffusion pathway for Na ions, and the introduced carbon coatings can significantly improve the electrical conductivity of the active materials. Increasing research attention, therefore, has been focused on this research field. For example, the pitch-derived carbon coating layers are introduced on the outer surface of the anatase TiO_2 nanorod, and this carbon modified TiO_2 showed a high reversible specific capacity of about 190 mAh g^{-1}. Moreover, due to the improved conductivity and reduced particle size, the rate performance was also significantly enhanced.[72] Apart from the pitch-derived carbon, the graphene also plays an important role in the TiO_2-based systems. The graphene-encapsulated rutile TiO_2 was created by Zhang's group, and this composite exhibited superior rate performance due to the graphene coating frameworks.[93] The cycling life and rate ability of the

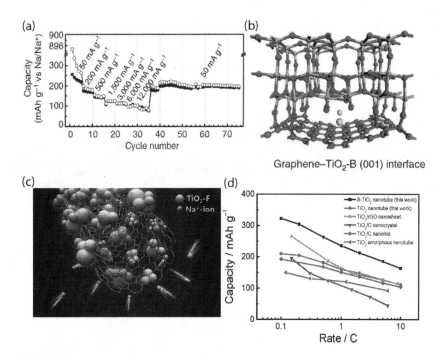

FIGURE 3.6
(a) Rate performance at various current densities from 50 to 12,000 mA g[−1] and (b) illustration of the partially bonded graphene-TiO$_2$-B (001) interface (Nature Publishing Group, Copyright 2015). (c) Schematic of F-doped TiO$_2$ embedded on CNTs (Copyright 2015, with permission from Elsevier). (d) Comparison of rate capability of S-TiO$_2$ nanotube arrays with other TiO$_2$-based nanostructured electrodes reported recently. The capacities were normalized at the same C rate (1 C = 335 mA g[−1]) (Copyright 2016 Wiley-VCH Verlag GmbH & Co. KGaA).

TiO$_2$-based anode can be further improved for the graphene-TiO$_2$ system if the extra chemical bond was introduced (Figure 3.6a and b).[91]

The heteroatom-doped TiO$_2$ also showed promising performance when used as an anode for SIBs. Hwang prepared the anatase TiO$_2$-carbon nanotube composite.[95] In this composite, the F-doped nanosized anatase TiO$_2$ particles were attached within the carbon nanotube (CNT) networks. Due to the introduction of this carbon nanotube framework, the conductivity of the TiO$_2$ system was improved, leading to the superior rate performance (Figure 3.6c). Furthermore, the F atoms contributed to the formation of the Ti^{3+} which was endowed with an outstanding electro-conducting ability, resulting in the effective and fast Na ions insertion into the TiO$_2$ crystalline structure. Bdoping is also a promising approach to improve the rate ability of TiO$_2$. According to the previous report, increasing number of Ti^{3+} and oxygen vacancies could be generated due to the Bdoping treatment, leading to the increased conductivity of TiO$_2$.[77] For example, Wang reported the B-doped TiO$_2$ last year using a scalable hydrothermal way and exhibited a high capacity of about 150 mAh g[−1] at a very high current density of 2C.[77]

Li's group prepared the sulfur-doped TiO_2 nanotubes, and this $S-TiO_2$ system demonstrated a high electrical conductivity (Figure 3.6d).[85] They explored the interaction effect between the introduced S and TiO_2. They found that part of the S 3p was delocalized after the doping process of S. Due to this unique ability, the width of the valence band was increased and the band gap energy was decreased.[53] Because of this reduced band gap energy, the conductivity and reversible capacity were both increased.

3.2.2.2 Lithium Titanate

Due to the flat and high working potential, the spinel $Li_4Ti_5O_{12}$ has attracted great research attention. Based on the previous study, $Li_4Ti_5O_{12}$ shows outstanding electrochemical performance, such as stable cycling life, when used as the anode in LIBs. Therefore, it is promising to use this material as the anode for SIBs because of the similar electrochemical reaction mechanism between SIBs and LIBs.

For LIBs, the spinel $Li_4Ti_5O_{12}$ is one of the most promising anode candidates for a long time due to its relevant working potential about 1.5 V (vs. Li/Li⁺) during lithiation/delithiation and outstanding cycling stability. According to reports over the last few years,' $Li_4Ti_5O_{12}$ is also a great anode candidate in SIBs. Chen reported that the spinel $Li_4Ti_5O_{12}$ could deliver a reversible capacity of 145 mAh g⁻¹ during the sodiation/desodiation process.[96] In addition, a low discharge voltage of about 1.0 V was also obtained during the sodiation stage. After the full sodiation process, a mixture of $LiNa_6Ti_5O_{12}$ and $Li_7Ti_5O_{12}$ was the main final outcome material. Based on this report, Huang proposed a further explanation about the sodiation behavior with the help of Density functional theory (DFT) calculations (Figure 3.7a–e).[56] Additionally, they also further improved the electrochemical performance of this $Li_4Ti_5O_{12}$ anode by adjusting the compositions of the binder and electrolytes.

Based on the previous relevant research about the $Li_4Ti_5O_{12}$-based LIBs, there is a two-phase transaction within the $Li_4Ti_5O_{12}$ anode during the sodiation process. When an increasing number of lithium ions are inserted into the $Li_4Ti_5O_{12}$, the crystalline structure of $Li_4Ti_5O_{12}$ will be changed from the spinel-typed structure into the rock-salt-typed $Li_4Ti_5O_{12}$ and rock-salt-typed $Li_7Ti_5O_{12}$ which is three more lithium ions than that of the $Li_4Ti_5O_{12}$. A different reaction was performed for the $Li_4Ti_5O_{12}$-based SIBs. For SIBs, there is a three-phase transformation that happens during the sodiation process, which is mainly due to different ionic size between Li ions and Na ions. Taking advantages of the ex situ techniques, Myung further explored the structural changes of Ti during the first cycling test (Figure 3.8a).[58] Yang also conducted a detailed study about this three-phase evolution of a $Li_4Ti_5O_{12}$-based anode for SIBs during the sodiation process.[340] Apart from the phase changes of a $Li_4Ti_5O_{12}$-based anode, the electrochemical behavior differences among different sized $Li_4Ti_5O_{12}$ electrodes were also studied by this group. They found that the reversible capacity of $Li_4Ti_5O_{12}$ was increased

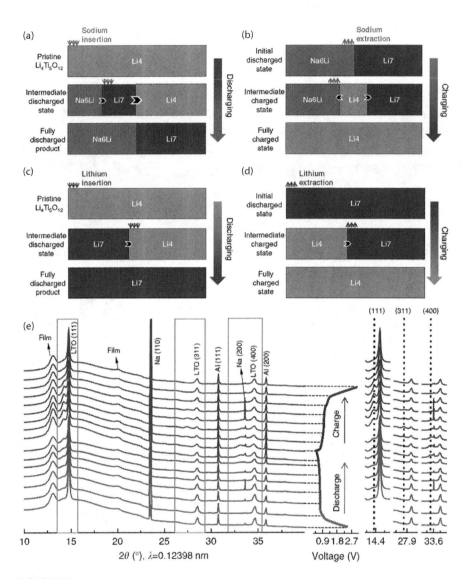

FIGURE 3.7
Comparison between sodiation and lithiation processes in Li$_4$Ti$_5$O$_{12}$. Discharging (a) and charging (b) processes in a sodium-ion battery. Discharging (c) and charging (d) processes in a lithium-ion battery. Li$_4$Ti$_5$O$_{12}$ (Li$_4$), Li$_7$Ti$_5$O$_{12}$ (Li$_7$), and Na$_6$LiTi$_5$O$_{12}$ (Na$_6$Li) phases are provided. Directions of phase boundary movement are marked by black arrows. (e) In situ synchrotron XRD patterns. The data were recorded during the first cycle for the Li$_4$Ti$_5$O$_{12}$ electrode in a sodium-ion battery system. For the Li$_4$/Li$_7$ phases, the main peaks correspond to (111), (311), and (400) reflections. These regions are highlighted in the right column and peaks corresponding to the Na6Li phase are marked by black dotted lines (Nature Publishing Group, Copyright 2012).

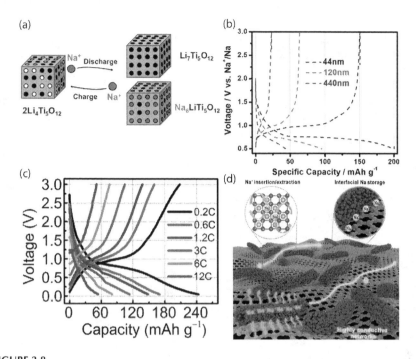

FIGURE 3.8
(a) Electrochemical reaction that occurred in carbon-coated Li$_4$Ti$_5$O$_{12}$ nanowires in the Na-cell (Copyright 2015, with permission from Elsevier). (b) The charge-discharge curve of Na storage into the crystallite size distribution of nanosized and submicrosized Li$_4$Ti$_5$O$_{12}$ (Copyright 2013 American Chemical Society). Sodium storage performance of the nanofibers wrapped with graphene (G-PLTO) electrode: (c) charge/discharge profiles from various C-rates of 0.2–12 C, and (d) graphical illustration of the structural merits and the integrated Na storage mechanisms in the G-PLTO electrode[2] (Copyright 2016 Wiley-VCH Verlag GmbH & Co. KGaA).

along with reducing the size of Li$_4$Ti$_5$O$_{12}$. For example, the reversible specific capacity of the 44 nm sized Li$_4$Ti$_5$O$_{12}$ was over ten times higher than that of the 440 nm sized Li$_4$Ti$_5$O$_{12}$ (Figure 3.8b). The increased capacity was mainly because of the shortened Na ions diffusion pathway during the sodiation/desodiation processes. A similar conclusion was obtained by Abe's group. They also found that reducing the size of Li$_4$Ti$_5$O$_{12}$ down to nanosize was an effective way to improve its rate performance and reversible capacity, because the Na ions sodiation/desodiation have a tight relationship with the inside architecture.[78]

Except reducing the particle size, introducing conductive coatings into the Li$_4$Ti$_5$O$_{12}$ material is another promising way to enhance its performance. Myung reported that the rate ability and reversible capacity (168 mAh g^{-1}) of Li$_4$Ti$_5$O$_{12}$ were both significantly improved by introducing the carbon coating layers which were derived from the pitch pyrolysis.[58] Huang's group prepared the Li$_4$Ti$_5$O$_{12}$ nanofibers with a porous structure inside via the electrospinning method and employed this carbon-Li$_4$Ti$_5$O$_{12}$ composite as the anode for the

SIBs (Figure 3.8c and d).[62] The Li$_4$Ti$_5$O$_{12}$ nanoparticles were coated by the carbon layers inside the carbon nanofibers and all the nanofibers connected with each other, forming a conductive interconnected framework, leading to outstanding electrical conductivity and structural stability.

3.2.2.3 Na$_2$Ti$_3$O$_7$

Compared with Li$_4$Ti$_5$O$_{12}$, Na$_2$Ti$_3$O$_7$ has attracted more research attention during the past few years because it has even lower working potential when used as an anode for SIBs.[38,40–43,45] This Na$_2$Ti$_3$O$_7$ material was first reported by Palacin's group in 2011. They reported that 2 Na ions can be reversibly inserted/extracted within one formula unit under a low operating voltage platform (0.3 V Na/Na$^+$)[38] (Figure 3.9a–d). This low working potential endowed

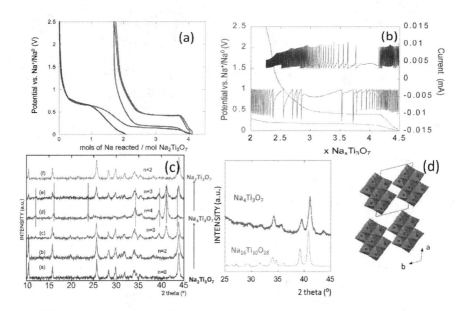

FIGURE 3.9
(a) Voltage versus composition profile for the electrochemical reduction of a blank electrode containing only carbon black and a composite electrode containing Na$_2$Ti$_3$O$_7$ and 30% carbon black where the reversible insertion of ca. 2 mol of sodium ions per mol Na$_2$Ti$_3$O$_7$ is observed. (b) Stepwise potentiodynamic experiment (5 mV steps with a cut-off intensity equivalent to C/150 galvanostatic rate). (c) *In situ* X-ray diffraction study of a Na$_2$Ti$_3$O$_7$/Na cell cycled between 2.5 and 0.01 V at a C/50 rate. For reasons of space, we have only reported the relevant XRD, the value of n corresponds to the mols of Na reacted per mol of Na$_2$Ti$_3$O$_7$. The similarity of patterns a and b are in agreement with the sodium uptake at 0.7 V being related to carbon. Further reduction induces the appearance of a new set of peaks (marked with a line) corresponding to a new phase, which grows with increasing sodium uptake and is pure at the end of reduction. (d) Experimental XRD pattern for Na$_4$Ti$_3$O$_7$ and simulated XRD pattern for Na$_{16}$Ti$_{10}$O$_{28}$ using the same profile parameters, where the analogy can be appreciated (left). Na$_{16}$Ti$_{10}$O$_{28}$ crystal structure (right) (Copyright 2011 American Chemical Society).

this material with more possibility to cooperate with suitable cathodes in one SIB system, leading to higher output energy density. An in-depth study about the sodium storage mechanism within the $Na_2Ti_3O_7$ anode was reported by Meng's group.[40] After the full sodiation process during the initial cycling test, 2 Na ions were inserted into the $Na_2Ti_3O_7$ structure, forming the $Na_4Ti_3O_7$ with a poor structural stability. At the same time, the chemical state of Ti was also changed from 4+ to 3+ along with the increasing number of Ti ions. The chemical state changes that were reversible happened between Ti^{3+} and Ti^{4+}, contributing to the reversible capacity of the $Na_2Ti_3O_7$. In order to overcome the poor structural stability of the newly formed $Na_4Ti_3O_7$, introducing carbon coating protective layers is a promising way.

3.3 Conversion Electrode Materials

Conversion reaction is another promising electrochemical reaction in the energy storage systems, because it can contribute high reversible specific capacity by adopting Na ions during the conversion process. The most promising candidates among all the conversion materials can be divided into three types, such as the transition metal sulfide (TMS)[97–140], transition metal oxide (TMO),[97,141–174] and transition metal phosphide (TMP).[175–187] Different from the alloying and intercalation reactions, in which the Na ions are reversibly inserted/extracted into/from the lattice of the active materials and a new compound will be generated after the sodiation process, the conversion reactions are also included for the conversion materials during the sodiation/desodiation process within the transition metals, leading to the high reversible specific capacity.[132] As a result, the conversion materials are promising as the potential candidates when used as the anode in LIBs and SIBs. Unfortunately, the large volume changes of conversion materials during the Na ions insertion/extraction process contribute to the poor structural stability, resulting in the poor cycling life of the conversion material-based energy storage systems. Moreover, the active materials will be peeled off from the current collector and mixed with the electrolyte during the cycling test when the volume changes happen, leading to the large capacity loss and poor conductive contact between the active materials and current collector. In addition, the larger size of the Na ion limits its efficient diffusion and transfer within the active materials. According to these disadvantages of conversion electrodes, various approaches have been proposed to improve the electrochemical performance of conversion materials, such as the design of the nanoarchitecture and introducing conductive coatings. Both reducing the size of the active particles and coating conductive layers can effectively improve the electrical conductivity and structural stability of the conversion materials, resulting in the stable cycling life and rate performance.

3.3.1 Transition Metal Oxides

The first report about the conversion electrode was proposed by Tirado in 2002. They found that the NiCo$_2$O$_4$ which is a conversion-typed material could be used as the anode material for SIBs (Figure 3.10a).[141] Based on this work, more reports have been published about the great number of TMOs, for example, iron/cobalt/tin/copper/molybdenum/nickel/manganese oxides.[158,174]

3.3.1.1 Iron Oxides

Yamada found that the sodiation reaction of Fe$_3$O$_4$ was conducted within the voltage range of 1.2–4.0 V when it was used as an anode for SIBs.[158] Balaya reported that the conversion reaction could also be obtained for the Fe$_3$O$_4$-based anode when the discharging potential was around 0.04 V.[142] After extending the discharging potential down to 0.04 V, the reversible capacity

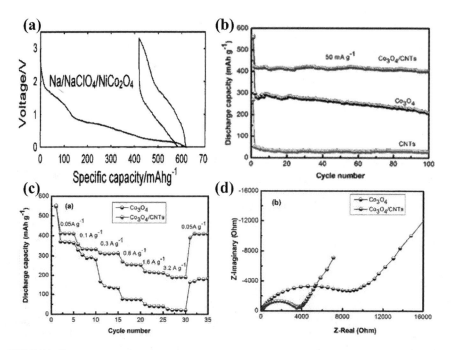

FIGURE 3.10
(a) Voltage profile of Na/NaClO$_4$ ethylene carbonate: dimethylcarbonate (EC:DMC)/NiCo$_2$O$_4$ cell between 3.3 and 0.01 V (Copyright 2002 American Chemical Society). (b) Cycling performance of CNTs, Co$_3$O$_4$, and Co$_3$O$_4$/CNTs electrodes. Consecutive cycling performance and electrochemical impedance spectroscopy of Co$_3$O$_4$ and composite Co$_3$O$_4$/CNTs electrodes: (c) consecutive cycling performance at different current rates, ranging from 0.05 to 3.2 A g^{-1}, followed by a return to 0.05 A g^{-1} and (d) electrochemical impedance spectroscopy of fresh cells (without cycling) (Copyright 2015 The Royal Society of Chemistry).

of the Fe₃O₄-based anode was increased to 643 mAh g⁻¹ during the first cycling test. However, the CE for this Fe₃O₄-based anode was low due to the large irreversible capacity loss during the initial discharge. The large irreversible capacity decay was mainly derived from the extra SEI films formation on the surface of the active particles. The large volume expansion of Fe₃O₄ produces fresh surface to be exposed to the electrolyte, leading to the continuous formation of SEI films. As a result, the active materials were irreversibly exhausted, resulting in the low CE. In order to overcome this problem, carbon coating layers are introduced into this system. After the introduction of carbon coatings, the structural stability of Fe₃O₄ will be significantly improved, and the stable SEI films will be generated on the outer surface. From this point of view, Myung created a CNTs/Fe₃O₄ composite which was composed of the interconnected CNTs, and the carbon coated Fe₃O₄ particles were uniformly embedded within this conductive CNTs framework.[159] As observed, a higher CE of 73% was obtained. In addition, the Fe₃O₄ quantum dots were also prepared by Qiu's group.[369] In this work, the Fe₃O₄ quantum dots were tightly attached on the carbon nanosheets, leading to a high conductive system. Similar to other nanosized materials, the reduced particle size can significantly improve their structural stability because the structural integrity of the Fe₃O₄ quantum dots could be well maintained even after the large volume expansion. Therefore, 70% of the capacity was well maintained even after 1000 cycling tests under the current density of 1.0 A g⁻¹.

3.3.1.2 Cobalt Oxides

Similar to Fe₃O₄-based anode materials, the Co₃O₄-based materials are also facing the same disadvantages when used as an anode for SIBs. According to the cyclic voltammogram technique, Chen's group explored the electrochemical reaction mechanism of Co₃O₄-based anode materials during the sodiation/desodiation process.[164] They found that there was a reversible conversion reaction for Co₃O₄-based SIBs: Co₃O₄ + 8Na⁺ + 8e⁻ ↔ 4Na₂O + 3Co. During the first discharge process, the Co₃O₄ was not fully converted into Na₂O, even though the discharging voltage was reduced to 0.01 V. In the following cycling test, the Co₃O₄ was gradually and continuously converted into Na₂O. As a result, the specific capacity of Co₃O₄-anode was increasing gradually in the first 20 cycles and reached the maximum 447 mAh g⁻¹ at the 20th cycling test. According to this report, Chen also created a CNT modified Co₃O₄-based composite to further improve the electrochemical performance of the Co₃O₄-based electrode.[148] For this work, the introduced CNTs played a crucial role for the final improved electrochemical performance, such as the increased electrical conductivity, shortened Na ions diffusion pathway, and enhanced structural stability. Therefore, the rate performance of this composite was significantly improved with a 190 mAh g⁻¹ at 3.2 A g⁻¹ (Figure 3.10b–d).

3.3.1.3 Tin Oxides

During the past 5 years, the nanosized tin (Sn)-based materials attracted great research attention, such as the tin-based oxide materials (SnO and SnO$_2$), because they showed promising ability when they were used as anode materials for SIBs.[150,151,170] The Sn-based materials also belong to the conversion-typed electrodes, because the Sn exhibited different chemical states during the reversible sodiation/desodiation reactions. Based on the *in-situ* TEM technique, Wang's group investigated the reaction mechanism of SnO$_2$ nanowires during the Na ions insertion/extraction stages when they were used as anode material for SIBs. In addition, they also studied the failure mechanism of this electrode to further improve the cycling stability (Figure 3.11a).[150] They found that a displacement-typed reaction happened during the first sodiation process of SnO$_2$, resulting in the disordered Na$_x$Sn surrounded by the Na$_2$O matrix. The crystallization degree of the Na ions-inserted Na$_x$Sn was also increased along with the further sodiation process, leading to the higher Na ions content inside the Na$_x$Sn ($3.75 \leq x \leq 15$). Compared with SnO$_2$, a large volume expansion was found for Na$_x$Sn when x reached 15. For example, the diameter was increased from 67 nm for the

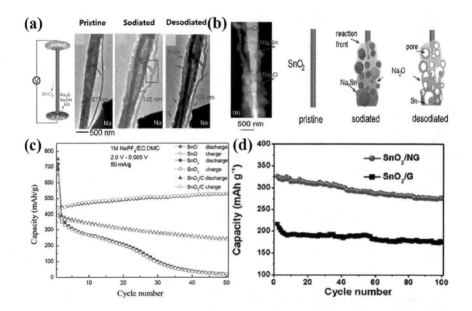

FIGURE 3.11
(a) TEM images of pristine, sodiated, and desodiated states of SnO$_2$ nanowire. (b) STEM Z-contrast image showing the reaction front of the SnO$_2$ nanowire (left) and schematic drawing showing the morphology evolution of the SnO$_2$ nanowire upon Na insertion and extraction (right) (Copyright 2013 American Chemical Society). (c) Cycle life of SnO, SnO$_2$, and SnO$_2$/C electrodes (Copyright 2015, with permission from Elsevier). (d) Cycling performance of SnO$_2$/NG and SnO$_2$/G composites at a current density of 20 mA g^{-1} (Copyright 2015 The Royal Society of Chemistry).

SnO$_2$ to 145 nm for the Na$_{15}$Sn. However, during the Na ions extraction stage, the Na$_{15}$Sn was converted into the Sn nanoparticles along with the generation of nanopores inside, resulting in the formation of a hollow Na$_2$O matrix which was embedded with the Sn nanoparticles. Due to the newly generated pores and void spaces, the electric impedance of this electrode was significantly increased, leading to the poor cycling stability and rate performance of SnO$_2$. The difference of the sodium storage ability between SnO and SnO$_2$ was also studied by Okada's group (Figure 3.11b).[168] And they found that the SnO with a lower oxygen percentage exhibited a higher reversible capacity and better rate performances, which means that the ratio between Sn and O is crucial to contribute capacity (Figure 3.11c). The electrochemical performance of SnO$_2$ can be optimized by the introduction of extra conductive carbon-based materials, such as the graphene, CNTs, and porous carbon.[168] Wang's group prepared the graphene embedded SnO$_2$ composite, and this nanohybrid showed a high capacity of about 340 mAh g^{-1} at the current density of 20 mA g^{-1} (Figure 3.11d).[153]

3.3.1.4 Copper Oxides

Because of the rich reserves, low-cost, and high theoretical reversible capacity of the copper-based oxides, the copper-based oxides are also the promising anode candidates for SIB, as are copper-based oxide materials.[154,173] For example, both Cu$_2$O and CuO exhibited high capacities of over 600 mAh g^{-1}.[171] Unfortunately, similar to other kinds of metal oxides, the copper-based oxides also have a large volume expansion during the sodiation process, leading to the structure damage and the following poor cycling stability. To address this large volume expansion, reducing the particle size and introducing the protection shell are promising ways because the reduced nanoparticles are endowed with enhanced structural stability during the volume changes. Chen's group reported that the carbon coated CuO nanoparticles showed a high reversible capacity of above 400 mAh g^{-1} after a long cycling test (600 cycles) under a current density of 200 mA g^{-1}.[156]

3.3.2 Transition Metal Sulfide

TMS is another promising anode candidate for SIBs because it is endowed with even higher theoretical capacity derived from the conversion reaction during the sodiation/desodiation process. Apart from the high theoretical capacity, compared with other traditional metal oxides, TMS materials also have other selling points due to the weaker M-S bonds and better reversibility of Na$_2$S which lead to the smaller volume expansion, higher initial CE, and great kinetic ability.[103] As a result, a great number of TMS materials, such as the CoS$_x$ (x = 1 and 2),[97,103] MoS$_2$,[104] FeS$_x$ (x = 1 and 2),[188] SnS$_x$ (x = 1 and 2),[119] CuS,[140] MnS,[126] NiS,[125] TiS$_2$,[189] WS$_2$,[128] and ZnS[129] have been studied as anode materials for SIBs.

3.3.2.1 CoS_x (x = 1 and 2)

Cobalt sulfide is a popular material and has been widely used in the semiconductor industry. During the past few years, the carbon modified CoS_x and nanostructured CoS_x showed promising ability to be employed as an anode in LIBs.[190] Fu's group reported the Na ions storage ability of the CoS_2-MWCNT composite and further studied the electrochemical performance of this CoS_2-Multi-walled carbon nanotube (MWCNT) composite within the carbonate-based or ether-based electrolytes (Figure 3.12a).[100] They found that the conversion reaction mechanism of this CoS_{x2}-MWCNT composite during the Na ions insertion/extraction process could be concluded as follows: $CoS_2 + 4Na^+ + 4e^- \leftrightarrow 2Co + 2Na_2S$. Compared with the pure CoS_2 anode, this CoS_2-MWCNT composite showed higher reversible capacity and enhanced CE (93%) due to the introduction of CNTs network which provided extra interconnected conductive frameworks, a fast electron/ion diffusion pathway, and larger surface area. After this work, Ramakrishna reported the electrochemical performance of the rGO modified CoS composite (rGO-CoS). They found that the rGO-CoS exhibited a high specific capacity (550 mAh g^{-1}) and outstanding cycling stability with a 88% capacity retention after 1000 cycling test.[191] Two years before, Cao's group prepared the polyaniline (PANI)-coated nanosized Co_3S_4 composite (PANI@Co_3S_4) and investigated the sodium storage ability of this material.[156] After the relevant electrochemical test was conducted, they found that the Na ions could be inserted/extracted into/from the PANI@Co_3S_4 composite via the conversion reaction.

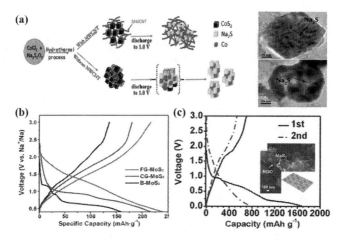

FIGURE 3.12
(a) Schematic diagram of the evolution process of the as-prepared CoS_2-MWCNT (left). And, ex situ TEM images of sodiated CoS_2-MWCNT and bare CoS_2 particles (right) (Copyright 2015 The Royal Society of Chemistry). (b) Charge and discharge curves of the as-prepared FG-MoS$_2$, CG-MoS$_2$, and B-MoS$_2$ at the first cycle (Copyright 2014 Wiley-VCH Verlag GmbH & Co. KGaA). (c) Galvanostatic charge-discharge profiles of MG-3 and SEM and TEM images of MG-3 (inset image) (Copyright 2015 Wiley-VCH Verlag GmbH & Co. KGaA).

3.3.2.2 MoS₂

For MoS$_2$, the covalent bonds could be found between the Mo and S atoms, leading to the 2D-typed S-Mo-S trilayers, and the adjacent planes within the MoS$_2$ are piled up by van der Waals interactions, leading to the fast Na ions insertion/extraction.[104] Under the working potential ranges, the Na ions can be inserted into the MoS$_2$ by the conversion reaction.[105] During the sodiation process, the theoretical reaction within the MoS$_2$-based anode in SIBs can be listed as follows: MoS$_2$ + xNa$^+$ + xe$^-$ → Na$_x$MoS$_2$ (over 0.4 V), Na$_x$MoS$_2$ + (4 − x)Na$^+$ + (4 − x) e$^-$ → 2Na$_2$S + Mo (below 0.4 V).[105]

Chen's group fabricated the interlayer distance of the (002) plane and increased the MoS$_2$ nanoflowers and reported that this material showed improved cycling life and rate performance through controlling the cut-off voltage ranges (Figure 3.12b).[105] After the electrochemical test, it can be found that this MoS$_2$ nanoflower exhibited stable cycling life and a high capacity retention rate during the 1500 cycling test. Unfortunately, the same disadvantages, such as the large volume expansion and low electrical conductivity, can also be found for the MoS$_2$ anode materials during the sodiation/desodiation stages. In order to overcome these drawbacks, Wang fabricated MoS$_2$ nanosheets with few-layered structure and found that the few-layer structured MoS$_2$ showed outstanding performance, such as the high reversible capacity (530 mAh g^{-1}) and superior rate performance.[111] Another layer-by-layer structured MoS$_2$/rGO composite was prepared by Wang's group and the electrochemical ability was also studied.[110] According to the DFT calculation and electrochemical test, they found that this unique layer-by-layer structure can provide extra active positions on the surface of MoS$_2$ nanosheets for Na ions insertion/extraction, resulting in the high reversible capacity of over 350 mAh g^{-1} at a very high current density (640 mA g^{-1}) within a working potential range of 0.01–3.0 V (Figure 3.12c).

3.3.2.3 FeS₂

FeS$_2$ is also a promising candidate as anode material for SIBs, because it has a high theoretical capacity of around 900 mAh g^{-1}. In addition, it is also a green material and less toxic to the environment. Kim studied the sodium storage ability of this FeS$_2$ material as an anode for SIBs for the first time.[114] The sodiation/desodiation process can be explained by the following reactions: FeS$_2$ + 2Na$^+$ + 2e$^-$ → Na$_2$FeS$_2$, Na$_2$FeS$_2$ + 2Na$^+$ + 2e$^-$ → 2Na$_2$S + Fe.[134] Due to the formation of Na$_2$S, the crystal structure of FeS$_2$ will be transferred to the amorphous structure, which mainly happens when the working potential is below 0.8 V. After the full sodiation process, FeS$_2$ nanoparticles delivered a reversible capacity of 500 mAh g^{-1} at 1A g^{-1}. Apart from the FeS$_2$ nanoparticles, FeS attracted less research attention due to its poor structural stability and conductivity.[192]

3.3.2.4 SnS$_x$ (x = 1 or 2)

There are two different kinds of tin-based sulfide materials, such as SnS and SnS$_2$. Both of these compounds are endowed with a high theoretical capacity.[120] For example, Cao's group reported that the SnS nanoparticles delivered a reversible capacity of 568 mAh g^{-1} at 20 mA g^{-1} after the carbon coating process.[119] More importantly, 98% of the reversible capacity was well maintained after 80 cycling tests of this SnS-carbon composite, indicating an outstanding structural and cycling stability. Yu's group reported that the interconnected porous frameworks of the pure SnS materials could also contribute high reversible capacity without the introduction of conductive additions.[136] Different from the crystal structure of SnS, SnS$_2$ is endowed with a unique sandwich structure which is composed of the covalently connected S-Sn-S layers.[137] Therefore, the interlayer distance among these layers is relatively large which makes it promising for the fast insertion/extraction of Na ions.[122]

3.3.3 Transition Metal Phosphide

Phosphorous materials also suffer from the same disadvantages when they are used as the anode material for SIBs, such as the large volume changes during sodiation/desodiation and poor conductivity. A promising way to address this problem is to create a binary metal-phosphide system via using the additional metal elements, such as Ni, Fe, Co, Cu, and Sn.[175] This can be explained by the newly generated intermediate compound, which happened between Na ions and the introduced metal elements during the Na ions extraction/insertion stages and can significantly improve the structural stability of the active materials.[183] For example, the Sn$_4$P$_3$ which was prepared by Lee's group delivered a high specific capacity over 700 mAh g^{-1}. More importantly, this sample showed outstanding cycling stability even after 100 cycling tests. The improved electrochemical performance of this sample was mainly because the structural damage of P and Sn during the alloy stage could be effectively suppressed by the conversion reaction process, leading to a unique self-healing ability (Figure 3.13a).[193] Apart from the pure Sn$_4$P$_3$ particles, additional structure design for the Sn$_4$P$_3$ is also another promising way to further improve its performance. For example, Yu's group fabricated the yolk-shell structured Sn$_4$P$_3$@Carbon anode material and showed great cycling stability.[179] The introduced void spaces between the inside Sn$_4$P$_3$ core and outside carbon shell can effectively accommodate the volume expansion of Sn$_4$P$_3$, which improved its structural stability. As a result, this yolk-shell structured Sn$_4$P$_3$@Carbon sample showed high capacity retention even after 500 cycling tests (Figure 3.13b). The self-healing process can be explained by Figure 3.13c.

Anode Materials for Sodium Ion Batteries

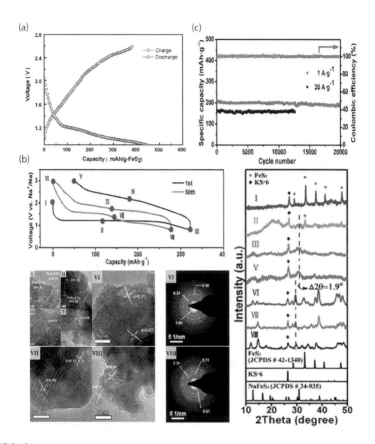

FIGURE 3.13
(a) Charge-discharge voltage profiles of the Na/FeS$_2$ cell at room temperature (Copyright 2008, with permission from Elsevier). (b) Structural evolution of FeS$_2$ during the electrochemical reaction with sodium: TEM images (left) and ex situ XRD results (right). (c) Cyclic performance of FeS$_2$ microspheres (Copyright 2015 The Royal Society of Chemistry).

3.4 Alloying Reaction Materials

Carbon-based and Ti-based oxide materials have been used as the anode materials for SIBs for a long time and showed outstanding cycling performance because of the small volume changes during the sodiation/desodiation process. Unfortunately, the relatively low theoretical capacity of these materials limited their practical utilization.

Apart from the traditional conversion materials, alloying-typed materials are also promising candidates when they are employed as anodes for SIBs, because the alloying-typed materials provide abundant active positions for Na ions insertion/extraction under a very low working voltage range.[194] Various kinds of interactions can be conducted with Na ions, leading to the high theoretical capacity. What kinds of alloying materials can be used as anode for SIBs? According to the previous research reports and publications, the elements such as Sn, Bi, Si, Ge, As, Sb, and P in the periodic table have great potential to be promising anode candidates for SIBs. However, there is also a large volume expansion during the sodiation process, which is mainly due to receiving a large amount of Na ions for these host materials. The repeated and continuous volume changes of the active materials during the cycling process can result in the structural pulverization. Furthermore, the damaged active materials will be separated from the current collector and diffused into the electrolyte, leading to the continuous capacity fading and safety risk.

3.4.1 Silicon, Germanium, Tin

3.4.1.1 Si

Si is the one of the most promising anode materials for LIBs because it has the largest theoretical capacity which is over 4000 mAh g^{-1} and abundant storage on the earth.[195] The large capacity of Si is mainly because Si can conduct the alloying reaction with Li ions in the formula of SiLi$_{4.4}$, leading to a very high specific capacity for Si anodes. As a result, Si is endowed with great potential as the anode candidate for commercial application. With increasing research attention focusing on SIBs, it is meaningful to explore the electrochemical performance of Si as an anode for SIBs. Yamane reported that Na ions can be alloy-typed to react with Si atoms, leading to the formation of the sodiated Na-Si.[196] However, Yamane also demonstrated that Si was not an available anode candidate for SIBs because it can only host one Na ion per Si atom, and it suffers from poor Na ions transfer kinetics (Figure 3.14a). Apart from that, Mulder's group experimentally studied the reversible alloying reaction between Si and Na ions (Figure 3.14a–d).[197] Mulder fabricated two different kinds of Si samples (the amorphous and crystalline Si particles). They found that the amorphous Si delivered a high reversible capacity of about 280 mAh g^{-1} and an outstanding cycling stability during 100 cycling tests. Lei Zhang also explored the electrochemically performance of crystal Si nanoparticles as anodes for SIBs. Lei studied the sodiation/desodiation ability of the nanosized Si nanoparticles, and they found that the crystal structure of Si was transferred to the amorphous structure after the first five cycling tests. In addition, they also found that the microsized Si particles are electrochemically inert for Na ions during the charge/discharge process due to the large particle size limited the diffusion of Na ions. Taking advantage of the *in-situ* Raman and XRD techniques, they successfully investigated the structural changes of Si during the cycling process. After the initial cycling test, the Si particles experienced an irreversible crystal structural change. The newly formed amorphous Si structure cannot turn back to its original crystal structure. Therefore, according to Lei Zhang's report, it can found that the electrochemical performance of crystalline Si

Anode Materials for Sodium Ion Batteries 51

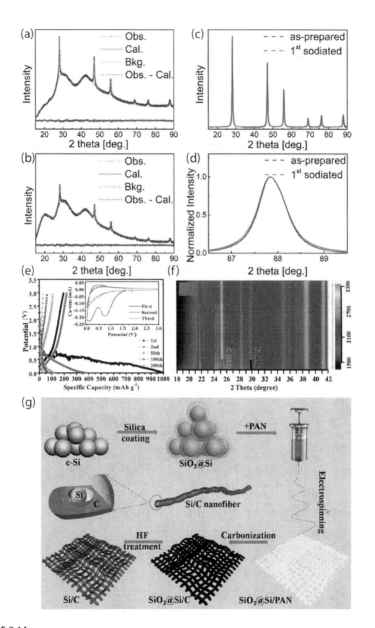

FIGURE 3.14
XRD patterns of the (a) as-prepared and (b) first cycle sodiated thick pellet electrode including Rietveld refinement. (c) Refined XRD patterns of an Si NP pressed pellet electrode before and after initial sodiation. (d) Lorentzian broadening (Lx) as observed at $2\theta = 87.9°$ before and after sodiation (Copyright 2015Wiley-VCH Verlag GmbH & Co. KGa). Electrochemical performance of the nanosized c-Si and microsized c-Si electrodes. (e) Charge-discharge profiles for selected cycles at 500 mA g^{-1} (inset: the first three Cyclic voltammetry (CV) curves for the nanosized c-Si electrode) and (f) *in operando* XRD data for microsized c-Si. (g) Schematic representation of the Si/C composite design (Copyright 2016 Wiley-VCH Verlag GmbH & Co. KGaA).

is tightly relying on its particle size (Figure 3.14e and f). According to their report, a large volume expansion also happened for Si anodes in SIBs which means similar problems, such as poor structural stability and cycling life, are also the disadvantages of Si-based anodes for SIBs. As a result, Lei Zhang fabricated the yolk-shell structured Si-based composite to further improve its electrochemical performance. The void space inside the yolk-shell structure can leave room for Si's volume expansion which means that the structure integrity of the Si-anode in SIBs can be better maintained (Figure 3.14g).[198,199]

3.4.1.2 Ge

Ge is similar to Si and is endowed with high theoretical capacity. Research attention has been attracted by Ge for a long time since its promising performance when used as an anode for LIBs. However, the commercial application of Ge-based anodes is also hindered by the same drawbacks as Si, such as the low electrical conductivity and large volume expansion during the charging/discharging process. Therefore, introducing conductive additions and void space into the Ge-based composites are crucial for its application.[200,201] The theoretical capacity of Ge is about 369 mAh g^{-1} based on its alloying reaction mechanism.[202] Similar approaches which have been successfully applied on the Si-based anodes can be directly used on Ge-based anode materials, such as reducing the particle size, introducing carbon coatings, and void space. For example, Veith's group fabricated the layered structured Ge films as anodes for SIBs and found that this Ge film delivered a reversible capacity of about 350 mAh g^{-1}.[203] In addition, they found that there was a two-phase transformation during the sodiation process with very flat charging or discharging voltage profiles.

3.4.1.3 Sn

Sn-based anode materials exhibit outstanding electrochemical performance with high theoretical capacity (around 850 mAh g^{-1}). After the full Na ion insertion process, $Na_{15}Sn_4$ is formed, leading to the high percentage of Na ions inside the Na-Sn alloy. Kuze's group reported that there was a reversible electrochemic al redox reaction conducted within the a Sn anode during the sodiation/desodiation process, leading to the reversible generation of the Sn-Na phase (Figure 3.15a).[204] Wang studied the crystal structure transformation and morphology changes of a Sn anode during the sodiation/desodiation process through the *in-situ* TEM technique (Figure 3.15b and c).[205] Based on their research, the sodiation process of Sn could be divided into two different stages, resulting in the formation of $NaSn_2$ after the first sodiation stage and Na_9Sn_4 under the second stage. After this report, based on the *in-situ* XRD technique, Wang's group experimentally studied the chemical/structural changes of a Sn anode during the charging/discharging process (Figure 3.15d and e).[206]

Anode Materials for Sodium Ion Batteries

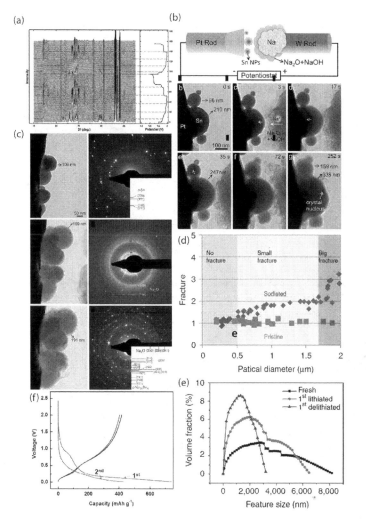

FIGURE 3.15
(a) *In situ* XRD data and the corresponding voltage curve. Dashed lines indicated the separating two-phase regions (Copyright 2012 The Electrochemical Society). (b) Experimental setup and the typical morphological evolution of Sn NPs during sodiation. (c) Microstructural evolution of Sn NPs during sodiation (Copyright 2014 American Chemical Society). (d) Statistical analysis of morphological complexity degree and (e) the feature size distribution of Sn particles in Sodium ion battery (NIBs) (Nature Publishing Group, Copyright 2015). (f) The capacity retention of the F-G/Sn@C electrode (Copyright 2016, with permission from Elsevier).

In order to further improve the cycling stability of Sn-based anode materials, a suitable structure design is necessary for Sn-based anodes. Usually, introducing yolk-shell, core-shell, and porous structures into the Sn-based composites is a promising approach to address this problem. Wang's group fabricated the free standing Sn-based electrode to improve its performance.[207] This unique free standing Sn-anode was composed of the pillar-like structured Sn/CNT composite, and these Sn/CNT nanopillars were grown on the surface of carbon paper, leading to a special Sn/CNT-CP composite. Due to the introduction of CNT and also the conductive carbon paper as the matrix, Sn/CNT-CP showed outstanding electrochemical performance, such as the high reversible capacity, superior rate performance, and improved cycling stability. The CNT and carbon paper accelerate the diffusion of Na ions and electrons within this conductive network. In addition, the nanopillar-like structure improved its structural stability during the large volume changes process. The ultra-small sized Sn composites also showed enhanced cycling life and rate performance. For example, the nanosized Sn particles which were mainly in the range of 1–8 nm in diameter were prepared by Liu's group.[208,209] In addition, Zhi's group also prepared the nanosized Sn particles with a typical size of about 15 nm and found that the nanosized Sn particles showed outstanding Na storage ability (Figure 3.15f).[210]

3.4.2 Antimony, Phosphorus, Bismuth, and Arsenic

Sb, P, Bi, and As are the elements which belong to the group 15 in the periodic table. Similar to Si, Ge, and Sn, these elements also exhibit promising performance via the alloying-typed reaction with Na ions when used as an anode for SIBs (Figure 3.16a–6d).[211] Due to the alloying reaction, large amounts of Na atoms can be reacted with these elements, leading to a high theoretical capacity. However, the same disadvantages, such as the large volume expansion after the full sodiation and low electrical conductivity, can also be found for the group 15 elements (Figure 3.16e and f). The large volume changes can cause serious damage to the structural integrity of the active materials, resulting in the poor cycling life and continuous waste of the electrolyte and active materials during the cycling test.

3.4.2.1 Sb

Antimony is also endowed with a high theoretical capacity of over 660 mAh g^{-1} based on the Na$_3$Sb which is the full Na ions inserted chemical formula.[212–214] Due to the large amount of Na ions within Na$_3$Sb, the sodiation reaction happens via a multi-step of an alloying/dealloying process. Yang's group first explored the sodiation/desodiation reaction mechanism. They found that a two-step reaction was conducted as follows: stage 1: Sb + Na$^+$ + e$^-$ ↔ NaSb, stage 2: NaSb + 2Na$^+$ + 2e$^-$ ↔ 2Na$_3$S (Figure 3.17a).[214] According to the *in-situ* XRD technique, Monconduit's group further studied in detail the real reaction process during the sodiation/desodiation process and found that an amorphous structured intermediate Na$_x$Sb phase was also generated (Figure 3.17b).[215] During the initial sodiation process, the Sb anode

Anode Materials for Sodium Ion Batteries 55

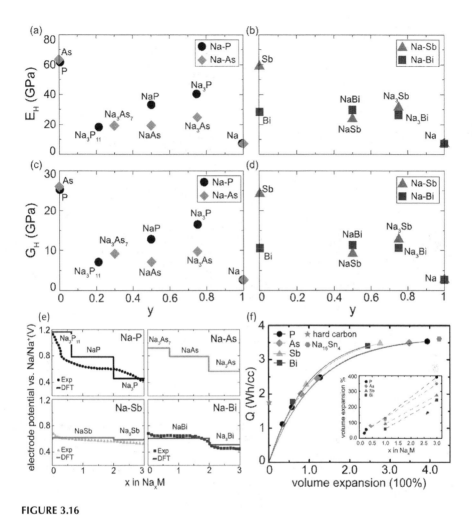

FIGURE 3.16
(a–d) Young's modulus (EH) and shear modulus (GH) of Nax M alloys as a function of Na fraction. (e) Calculated electrode potential profiles with respect to Na/Na+ for Na_xM phases and (f) the volumetric energy density as a function of volume expansion for Na_xM phases. (M = P, As, Sb, and Bi) (Copyright 2015, with permission from Elsevier).

was initially changed to an intermediated phase (Na_xSb) with disordered crystalline structure. After the sodiation reaction was mostly finished, the amorphous Na_xSb was shifted to the crystalline Na_3Sb with a cubic-hexagonal structure. In the following Na ions extraction stage, the Na_3S was converted back to the amorphous structured Sb. Due to the formation of the amorphous Na_xSb, the cycling stability of Sb was significantly improved due to this Na_xSb buffer to accommodate the produced strain and maintain the structural stability of Sb. Unfortunately, due to the large number of Na atoms

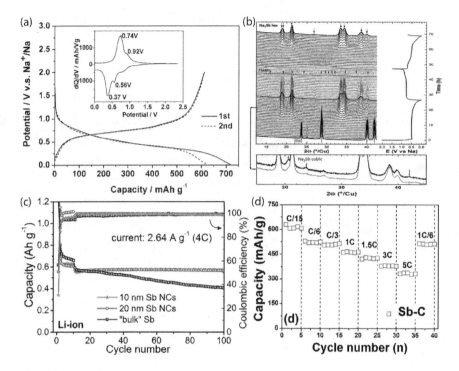

FIGURE 3.17
(a) Electrochemical properties of the Sb/C nanocomposite: charge-discharge curves of the Sb/C composite at a constant current of 100 mA g^{-1}, the inset is differential capacity versus cell potential curve (Copyright 2012 The Royal Society of Chemistry). (b) Operando evolution of the XRD pattern recorded at a C/8 rate (top left). The patterns are those recorded during the discharges and charge, respectively. The corresponding voltage profile (top right). A zoom illustrating the diffraction peaks from the cubic Na$_3$Sb (bottom) (Copyright 2012 American Chemical Society) (c) Cycling performance of Sb electrodes with different size distribution in Na ions half-cells. (Copyright 2014 American Chemical Society). (d) C rate capability of the Sb-C electrode at various current rates from C/15 to 5C (Copyright 2014 The Royal Society of Chemistry).

within the Na-Sb alloy which contributes the high theoretical capacity, there is also a large volume expansion after the full sodiation process, leading to a 400% volume expansion of Sb anodes.[216] As we mentioned before, the large volume changes of the electrodes damage the structural integrity of the active materials and further result in poor cycling life. Therefore, the research focus on Sb-based anodes is focusing on the structural design to relieve the structure damage of Sb, such as reducing the particle size from microsize to nanosize, introducing a protection shell on the outer surface of Sb, and creating void space inside the active materials.[216–226] Downsizing of the Sb particles is one of the most promising ways to address the poor structural stability, because it can effectively accelerate the electron transfer, benefit the kinetics, and give better ability to stand the volume changes. Kovalenko's group prepared the monodispersed Sb nanocrystals with a typical size in the range of 10–20 nm and further employed this nanocrystal structured Sb as an anode for SIBs.[218] Compared with the conventional microsized the Sb anodes, these downsizing Sb nanocrystals showed significantly improved electrochemical performance in the half SIBs (Figure 3.17c). Apart from reducing the particle size, other groups explored another way to improve Sb-based anode's performance by introducing the conductive matrix to enhance its conductivity and cycling stability. For example, Cao's group fabricated the Sb nanoparticle embedded carbon nanofibers in 2014. The created composite exhibited a high capacity which was over 630 mAh g^{-1} with outstanding reversible capacity retention (over 400 cycles) at a higher current density (Figure 3.17d). Bao's group introduced graphene into the Sb-based composite systems, in which the Sb nanoparticles were uniformly embedded on the graphene layers.[226] The biggest problem for carbon modified Sb-based anodes is the weak interaction between the carbon holder and Sb particles, leading to the poor structural stability of the composite. In order to enhance the overall stability, a strong chemical bond was introduced into this system to strengthen the interconnection between Sb and the carbon matrix. The strong chemical bonded Sb-C composites are endowed with increased electrical conductivity and improved structural stability, leading to the superior rate and cycling performance. For example, the chemical bonded graphene-Sb composite delivered a high capacity over 450 mAh g^{-1} and promising cycling stability with 90% capacity retention over 200 cycles.

3.4.2.2 P

Due to the light weight of Na$_3$P and reversible reaction between Na and P, P as a promising anode candidate, attracted great research attention. A highest theoretical capacity, which is as high as 2600 mAh g^{-1} can be obtained for P-based anodes for SIBs.[227–229] There are three different types of P materials, red P, black P, and white P (Figure 3.18a).[227,229–232] Compared with red and black P, white P is the most active one which can be burned into flames under the air atmosphere. However, red P and black P are relatively stable in the air. Therefore, all the research about P-based anodes is focusing on the red and black P-based anode materials.

As for the red and black P-based materials, the poor electrical conductivity and large volume expansion (490%) after sodiation are the most important disadvantages that hinder the further application of P-based electrodes.[230] Yang's group reported that the carbon embedded red-P composite showed promising performance when used as an anode for SIBs (Figure 3.18b).[231] During the Na ions insertion process, the red P delivered a very high capacity of around 900 mAh g^{-1}. However, the reversible charging capacity was only about 15 mAh g^{-1} during the following charging process, demonstrating that this red P/C material was inactive for sodiation due to its

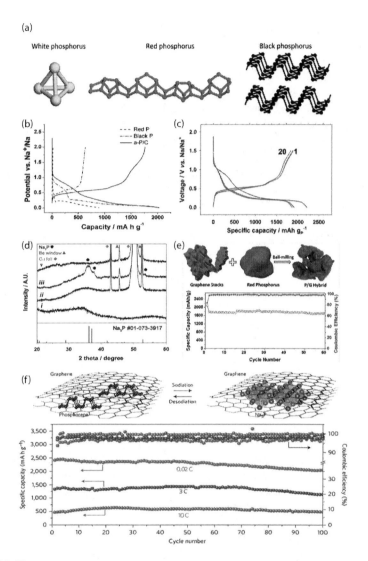

FIGURE 3.18
(a) Schematics of three different types of phosphorus (Copyright 2014 American Chemical Society). (b) Voltage profiles of the a-P/C sample charged at a current density of 250 mA g^{-1} and then discharged at a different current density from 250 to 4000 mA g^{-1} (Copyright 2013 Wiley-VCH Verlag GmbH & Co. KGaA). (c) Voltage profiles (black) of amorphous red P/C composite and the corresponding electrode thickness change (gray) during sodiation and desodiation. (d) Ex situ XRD patterns of amorphous red P/C composite electrodes collected at various points as indicated in (c) (Copyright 2013 Wiley-VCH Verlag GmbH & Co. KGaA). (e) Rate capability of the P/G hybrid anode (Copyright 2014 American Chemical Society). (f) Reversible desodation capacity and Coulombic efficiency for the first 100 galvanostatic cycles of the phosphorene/graphene (48.3 wt% P) anode tested under different currents (Nature Publishing Group, Copyright 2015).

insulating electronic ability. On the other hand, the amorphous red P material exhibited greatly improved electrochemical performance compared with the crystalline red P. For example, Lee's group reported that the amorphous red-P/C composite showed a high reversible capacity of about 1540 mAh g^{-1} at a high current density (Figure 3.18c).[227] Taking advantages of the *in-situ* XRD technique, the Na$_3$P phase was also found during the sodiation process (Figure 3.18d). The nanosized amorphous red P-based composites embedded within the 2D or 3D carbon holders were prepared to address the poor structural stability.[231] Wang's group fabricated the red P/graphene composite via the scalable ball milling method (Figure 3.18e).[230] During the ball milling process, the graphene layers were exfoliated and the strong chemical bonds were generated between the graphene layers and nanosized red P particles, leading to the red P nanoparticle embedded graphene composite. Due to the introduction of the chemical interaction between red P and graphene, the conductivity of this composite was significantly improved. In addition, the exfoliated graphene layers contained rich void space among different graphene layers, which could better accommodate the volume changes of red P particles during the cycling test, resulting in the outstanding cycling life. Therefore, this red P/graphene exhibited a high reversible capacity of about 2080 mAh g^{-1} and good cycling stability of 1700 mAh g^{-1} after 60 cycles.

Compared with red P, black P is endowed with even higher electrical conductivity, better structural stability, and a higher crystalline degree. The black P has a layered structure which is similar to that of the graphite, but the interlayer distance is larger than graphite.[230] The increased interlayer distance and higher conductivity make black P more promising as the anode for SIBs.[231] Glushenkov's group fabricated the graphite/black P nanoparticle composite through the ball milling way and explored the electrochemical performance of this sample at various cut-off voltage windows.[233] They found that this composite could contribute a very high initial capacity of 1300 mAh g^{-1} in the working potential range of 0.01–2 V versus Na/Na$^+$. Unfortunately, this electrode suffered serious capacity loss in the following cycling test. A different performance could be obtained when the voltage window was changed to 0.33–2 V versus Na/Na$^+$. Effectively improved cycling stability over 100 cycles was achieved for the same electrode material under this new voltage range. Based on the SEM images of the electrode after the cycling test under different voltage ranges, it could be found that the active material was separated from the current collector and dissolved into the electrolyte within a wider voltage window (0.01–2 V vs. Na/Na$^+$). However, this problem could not be found for the cell which was tested under 0.33–2 V versus Na/Na$^+$, resulting in the more stable cycling life. In order to obtain a better performance of a black P-based anode in SIBs, Yi Cui's group proposed a novel design which was composed of a unique sandwich structured black P/graphene composite. The few layered black P nanosheets were sandwiched between graphene layers (Figure 3.18f).[229] This sandwich-like P/graphene composite is endowed with several selling points: (1) the flexible graphene layers provide the buffer to absorb the volume changes from black P during the charging/discharging process, (2) the few-layered structured black P nanosheets have enlarged interlayer distance which can offer a reduced pathway for sodiation/desodiation, and (3) the chemical bonds between the black P and graphene surface provide effective electrical highway. Therefore, a high specific capacity of over 2400 mAh g^{-1} within 0.33–2 V versus Na/Na$^+$ and promising capacity retention during 100 cycling tests were both obtained for this unique black P/graphene composite. To better understand the real reaction mechanism of black P anode during the sodiation/desodiation process, Komaba's group studied the crystalline structure changes of black P with the help of *ex situ* XRD.[234] Before the sodiation stage, the black P has a crystalline structure with the orthorhombic lattice inside. After the full Na ions insertion, orthorhombic-typed black P particles were converted to the hexagonal phased Na$_3$P, which is the same as the red P. During the following desodiation process, the crystalline structured black P was not found which means the crystalline structure changes during the charging/discharging were not reversible. Therefore, it can be concluded that the crystalline structured black P is endowed with the metastable ability, and disordered P could be obtained in the form of the oxidation product within the half SIBs.[234]

3.4.3 Binary Inter-metallic Compounds

Designing the binary inter-metallic alloy materials is another promising way to improve the electrochemical performance of the anode materials due to the novel physiochemical properties. The Sn-M- and Sb-M-typed binary materials (M: metal elements) are the hot research topics among all the different binary compounds. Based on the difference of the electrochemical activity of the M elements, these binary materials can be divided into two main categories: electrochemical inactive ones and active ones. The inactive M elements are mainly including the Ni, Zn, Fe, Mo, and Cu,[235–240] while the active ones include the Bi, Sn, and Sb which conduct the alloying-dealloying reaction with Na ions during the sodiation/desodiation process within the SIBs.[230,241–243] For the active M-based binary materials, there is a two-reaction mechanism that happened during the Na ions insertion/desertion. Different from the Sn or Sb, the introduced secondary M elements contribute to the cycling stability of the binary materials. As we mentioned before, there is an intermediate phase generated with amorphous structure during the sodiation/desodiation process of Sn or Sb anodes. With the introduction of the secondary M element, another intermediate phase derived from the second introduced M elements can also be formed during cycling, which means there will be double intermediate phases generated within the active materials. The generated double intermediate phases can act as a role of buffer to more effectively absorb the volume changes of the electrodes during Na ions insertion/extraction, leading to the improved structural stability and outstanding cycling life.[243,244]

Yu's group fabricated the Sn nanoparticles embedded porous-structured microsized Ni$_3$Sn$_2$ cages (Figure 3.19a–c).[235] This Ni-Sn compound was

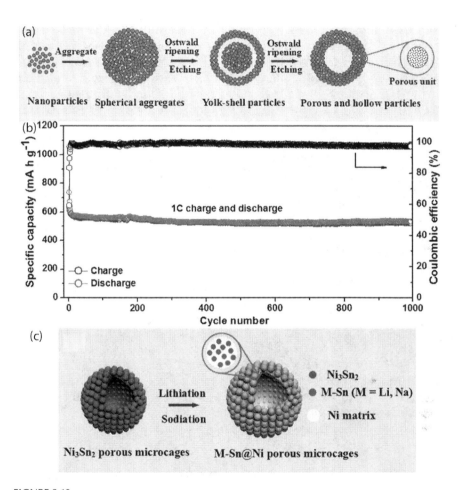

FIGURE 3.19
(a) Schematic illustration of the formation process of hollow-structured Ni$_3$Sn$_2$ porous microcages based on Ostwald ripening and etching. (b) Charge-discharge capacities of Ni$_3$Sn$_2$ porous microcages for lithium batteries over 1000 cycles at 1C and (c) schematic representation of the first lithiation and sodiation of porous Ni$_3$Sn$_2$ inter-metallic microcages, forming 0 D electroactive M–Sn (M = Li, Na) particles embedded in 3D conducting Ni hollow matrix (Copyright 2014 American Chemical Society.)

employed as the anode for SIBs and followed the double alloying-dealloying reaction mechanism as follows: $Ni_3Sn_2 + 7.5\ Na^+ + 7.5e^- \rightarrow 2Na_{3.75}Sn + 3Ni$, $Na_{3.75}Sn \rightarrow Sn + 3.75Na^+ + 3.75e^-$. After the Na ions fully inserted into the Ni_3Sn_2 electrode, the porous-structured Ni_3Sn_2 cages were turned into the Na-Sn nanoparticles with 0 dimensional structure and a 3D-structured hollow matrix which was embedded with Ni particles. The void space inside this material was crucial to improve its structural stability during the cycling test and provided extra active positions for the Na ions insertion/extraction. As a result, this anode electrode showed high reversible capacity of about 350 mAh g^{-1} with high capacity retention over 90% during a long cycling test (300 cycles).

Another binary compound, Sb-based binary materials which include Cu_2Sb and $FeSb_2$ compounds, also attracted great research attention. During the initial Na ions insertion process, the crystalline-structured Na_3Sb nanocrystal particles were first generated which were derived from the reaction between Sb and Na ions. After that, the disordered structured Na-Cu-Sb and Fe_4Sb were formed along with further Na ions insertion. Unfortunately, during the sodiation process, part of the reactions to produce these intermediate compounds were irreversible which means these parts of the intermediate materials could not be turned back to the original chemical states, leading to the active material waste and continuous capacity lose.

3.5 Organic Compounds

Compared with the inorganic electrode materials, the organic ones attracted less research attention because the inorganic electrodes have been widely researched for a long history and are already commercially available for practical application.[245] Unfortunately, for some special applications, such as the flexible batteries with reduced cost and non-toxic ability, the organic electrodes are necessarily required during the battery's fabrication. The main reasons for that are because of the advantages of organic electrodes, such as their mechanical flexibility, large stable structure changes, low density, low-cost, environmentally friendly ability, and chemical diversity, which provides a promising application of organic materials as a candidate for special batteries.[246,247]

Among all the organic materials, the carbon salts are the most promising candidates when they are used as anodes for energy storage systems, because their structure can be tailored with the design and selection of organic parts and metal ions. Different from the inorganic materials with large volume expansion and poor cycling stability, the organic electrodes have better structure stability because the volume and structure of organic carbonyl groups are less affected by the Na ions insertion/extraction due to their flexible ability, leading to the improved electrochemical performance.[245]

The organic carbonyl materials which are available to be used as anodes for SIBs can be mainly divided into three different types, including the conjugated carboxylates-, Schiff base, and imides and quinones-groups. Although the organic anodes have advantages, their commercial application is still hindered by some drawbacks, including the poor reaction kinetics which are derived from the poor electrical conductivity and chemical instability when the organic compounds are contacted with organic electrolytes.[245] Both the low conductivity and the active reaction between the organic compounds and electrolytes can lead to serious capacity loss, low initial Coulombic efficiency, and poor cycling stability. In the following section, the sodium storage ability of different organic materials when used as anodes will be discussed.

The conjugated carboxylates functional groups-based materials exhibit the redox-typed reaction with reversible ability under low operation voltage ranges (0.2–0.5 V vs. Na$^+$/Na) and great reversible sodiation/desodiation ability over long cycling tests.[245,248-255] The terephthalate-based materials are also great potential anode electrodes for SIBs.[235] Chen's group reported the Na$_2$C$_8$H$_4$O$_4$/ketjen black (KB) composite as the anode material in SIBs for the first time and found that this novel anode showed an outstanding cycling stability and a high capacity of 250 mAh g^{-1} (Figure 3.20a and b).[246] In addition, the Na$_2$C$_8$H$_4$O$_4$/KB composite with an atom level Al$_2$O$_3$ coating layer has also been employed as an anode for SIBs, resulting in outstanding electrochemical performance. Different kinds of disodium terephthalate derivatives, for example, the NH$_2$-Na$_2$TP, Br-Na$_2$TP, and NO$_2$-Na$_2$TP, were also studied and reported as anodes for SIBs by Hong's group during the past few years.[247] According to their research, the Na$_2$TP-typed anode showed a high capacity of around 300 mAh g^{-1} after full sodiation/desodiation. After the Na ions insertion, the phenyl rings and carbonyl functional groups can be conjugated together, leading to the formation of the more stable structure. Compared with Na$_2$TP-typed anodes, the NO$_2$-Na$_2$TP- and Br-Na$_2$TP-based anodes exhibited higher reversible capacities of over 300 mAh g^{-1}, but the NH$_2$-Na$_2$TP compounds showed a lower specific capacity of about 200 mAh g^{-1}. Amine's group also explored the sodium storage ability of disodium terephthalate-typed compound for the first time (Figure 3.20c and d).[248] They found that there was a four-electron transfer during the fully sodiation/desodiation process of the Na$_4$C$_8$H$_2$O$_6$ electrode. Na$_4$C$_8$H$_2$O$_6$ can be inserted with two more Na ions and also extracted with two more Na ions

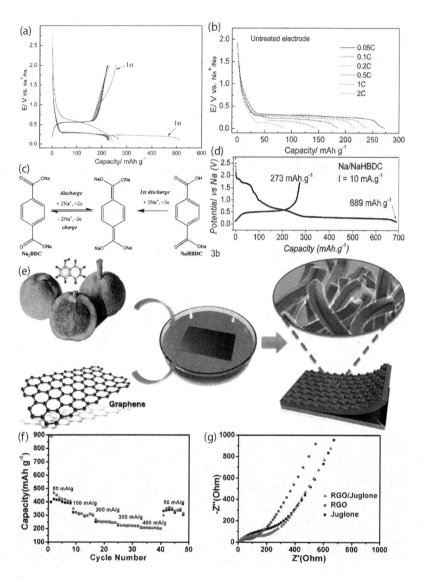

FIGURE 3.20
The first ten discharge-charge cycles (a) and the discharge curves (b) of the untreated Na$_2$C$_8$H$_4$O$_4$ electrode (Copyright 2012 Wiley-VCH Verlag GmbH & Co. KGaA). (c) Representative charge-discharge process on Na$_2$BDC and Nahybrid converter topologies, and (d) the initial discharge-charge curves of Na/NaHBDC cell (Copyright 2012 The Royal Society of Chemistry). (e) Fabrication process of Juglone/rGO electrodes, and their rate capability (f) and electrochemical impedance spectra (g) (Copyright 2015 Wiley-VCH Verlag GmbH & Co. KGaA).

as well, leading to the formation of $Na_6C_8H_2O_6$ and $Na_2C_8H_2O_6$, respectively. As a result, this $Na_4C_8H_2O_6$ material can be treated as both the cathode and anode in one energy storage system, resulting in the possibility of this electrode to be employed as the symmetric electrode in the full SIBs. Based on their study, two Na ions can be extracted from $Na_4C_8H_2O_6$ at 2.3 V and another two Na ions are inserted into $Na_4C_8H_2O_6$ at 0.3 V, leading to a 2.0 V voltage gap between the charging and discharging stages.[256] Taking adva1ntages of this unique ability, the full symmetric SIBs with the $Na_4C_8H_2O_6$ as both the anode and cathode were fabricated. This all organic symmetric battery exhibited high capacity of about 180 mAh g^{-1} and showed a high working potential of around 1.8 V. In order to improve the Na ions diffusion and transfer within the electrode, Lei's group fabricated the sodium 4,40-stilbene-dicarboxylate (SSDC) as the organic anode for SIBs.[254] They found that this SSDC-based anode delivered a reversible capacity over 100 mAh g^{-1} at a very high current density (2 A g^{-1}) and 72 mAh g^{-1} at an even higher current density of 10 A g^{-1}. Based on this research, the improved electrochemical performance of a SSDC-based electrode can be mainly supported by the stable charge transport, the increased inter-molecular reactions. However, the retention of the capacity during the long cycling test was poor due to the dissolution of this SSDC within the polar-typed electrolyte.[257] To better improve the cycling stability, the continuous dissolution of active materials which are coated on the current collectors should be suppressed. The promising way to address this problem is to introduce the protection layers on the surface of the active electrodes, such as the polymer coatings.[251,257–259] Xia's group introduced the 1, 4, 5, 8-naphthalenetetracarboxylic dianhydride derived polyimide (PNTCDA) which is a polyimide-based polymer as the protection layer to further suppress the dissolution of the organic anode materials.[259] As a result, the cycling stability and initial Coulombic efficiency (96.7%) of this modified anode was significantly improved due to the protection derived from the coating layers.

Apart from the abovementioned organic materials, the biomolecule-based materials are also promising selections to be used as anodes for SIBs. For these biomolecule-based organic materials, the quinine and carbonyl groups are included.[260–264] Chen's group reported that the juglone-typed biomolecule-based organic compound showed outstanding electrochemical performance when used as an anode for SIBs, because the redox quinine carbonyl functional groups inside (Figure 3.20e and f).[261] Furthermore, the juglone also showed further improved performance when embedded on the surface of rGO sheets due to the strong chemical interaction between the graphene and aromatic functional groups. Therefore, this composite showed a high reversible capacity of over 300 mAh g^{-1} and outstanding cycling life.

References

1. Aurbach, D., Levi, M. D., Levi, E., Teller, H., Markovsky, B., Salitra, G., Heider, U., Heider, L. Common electroanalytical behavior of Li intercalation processes into graphite and transition metal oxides. *J. Electrochem. Soc.* **1998**, *145* (9), 3024–3034.
2. Reynier, Y. F., Yazami, R., Fultz, B. Thermodynamics of lithium intercalation into graphites and disordered carbons. *J. Electrochem. Soc.* **2004**, *151* (3), A422–A426.
3. Pascal, G. E., Fouletier, M. Electrochemical intercalation of sodium in graphite. *Solid State Ion.* **1988**, *28*, 1172–1175.
4. Raccichini, R., Varzi, A., Passerini, S., Scrosati, B. The role of graphene for electrochemical energy storage. *Nat. Mater.* **2015**, *14* (3), 271–279.
5. Wang, Z. H., Selbach, S. M., Grande, T. Van der Waals density functional study of the energetics of alkali metal intercalation in graphite. *RSC Adv.* **2014**, *4* (8), 4069–4079.
6. Jache, B., Adelhelm, P. Use of graphite as a highly reversible electrode with superior cycle life for sodium-ion batteries by making use of co-intercalation phenomena. *Angew. Chem. Int. Edit.* **2014**, *53* (38), 10169–10173.
7. Zhu, Z. Q., Cheng, F. Y., Hu, Z., Niu, Z. Q., Chen, J. Highly stable and ultrafast electrode reaction of graphite for sodium ion batteries. *J. Power Sources* **2015**, *293*, 626–634.
8. Jache, B., Binder, J. O., Abe, T., Adelhelm, P. A comparative study on the impact of different glymes and their derivatives as electrolyte solvents for graphite co-intercalation electrodes in lithium-ion and sodium-ion batteries. *Phys. Chem. Chem. Phys.* **2016**, *18* (21), 14299–14316.
9. Kim, H., Hong, J., Yoon, G., Kim, H., Park, K. Y., Park, M. S., Yoon, W. S., Kang, K. Sodium intercalation chemistry in graphite. *Energ. Environ. Sci.* **2015**, *8* (10), 2963–2969.
10. Kim, H., Hong, J., Park, Y. U., Kim, J., Hwang, I., Kang, K. Sodium storage behavior in natural graphite using ether-based electrolyte systems. *Adv. Funct. Mater.* **2015**, *25* (4), 534–541.
11. Wen, Y., He, K., Zhu, Y. J., Han, F. D., Xu, Y. H., Matsuda, I., Ishii, Y., Cumings, J., Wang, C. S. Expanded graphite as superior anode for sodium-ion batteries. *Nat. Commun.* **2014**, *5*, 4033.
12. Doeff, M. M., Ma, Y. P., Visco, S. J., Dejonghe, L. C. Electrochemical insertion of sodium into carbon. *J. Electrochem. Soc.* **1993**, *140* (12), L169–L170.
13. Stevens, D. A., Dahn, J. R. High capacity anode materials for rechargeable sodium-ion batteries. *J. Electrochem. Soc.* **2000**, *147* (4), 1271–1273.
14. Irisarri, E., Ponrouch, A., Palacin, M. R. Review-hard carbon negative electrode materials for sodium-ion batteries. *J. Electrochem. Soc.* **2015**, *162* (14), A2476–A2482.
15. Komaba, S., Murata, W., Ishikawa, T., Yabuuchi, N., Ozeki, T., Nakayama, T., Ogata, A., Gotoh, K., Fujiwara, K. Electrochemical Na insertion and solid electrolyte interphase for hard-carbon electrodes and application to Na-ion batteries. *Adv. Funct. Mater.* **2011**, *21* (20), 3859–3867.
16. Wenzel, S., Hara, T., Janek, J., Adelhelm, P. Room-temperature sodium-ion batteries: Improving the rate capability of carbon anode materials by templating strategies. *Energ. Environ. Sci.* **2011**, *4* (9), 3342–3345.
17. Tsai, P. C., Chung, S. C., Lin, S. K., Yamada, A. Ab initio study of sodium intercalation into disordered carbon. *J. Mater. Chem. A* **2015**, *3* (18), 9763–9768.

18. Bommier, C., Surta, T. W., Dolgos, M., Ji, X. L. New mechanistic insights on Na-ion storage in nongraphitizable carbon. *Nano Lett.* **2015**, *15* (9), 5888–5892.
19. Berger, C., Song, Z. M., Li, X. B., Wu, X. S., Brown, N., Naud, C., Mayou, D. et al. Electronic confinement and coherence in patterned epitaxial graphene. *Science* **2006**, *312* (5777), 1191–1196.
20. Liu, J. H., Liu, X. W. Two-dimensional nanoarchitectures for lithium storage. *Adv. Mater.* **2012**, *24* (30), 4097–4111.
21. Zhu, J. X., Yang, D., Yin, Z. Y., Yan, Q. Y., Zhang, H. Graphene and graphene-based materials for energy storage applications. *Small* **2014**, *10* (17), 3480–3498.
22. Wang, Y. X., Chou, S. L., Liu, H. K., Dou, S. X. Reduced graphene oxide with superior cycling stability and rate capability for sodium storage. *Carbon* **2013**, *57*, 202–208.
23. Yan, D., Xu, X. T., Lu, T., Hu, B. W., Chua, D. H. C., Pan, L. K. Reduced graphene oxide/carbon nanotubes sponge: A new high capacity and long life anode material for sodium-ion batteries. *J. Power Sources* **2016**, *316*, 132–138.
24. Sun, Y. G., Tang, J., Zhang, K., Yuan, J. S., Jing, L. A., Zhu, D. M., Ozawa, K., Qin, L. C. Comparison of reduction products from graphite oxide and graphene oxide for anode applications in lithium-ion batteries and sodium-ion batteries. *Nanoscale* **2017**, *9* (7), 2585–2595.
25. Yabuuchi, N., Kubota, K., Dahbi, M., Komaba, S. Research development on sodium-ion batteries. *Chem. Rev.* **2014**, *114* (23), 11636–11682.
26. David, L., Singh, G. Reduced graphene oxide paper electrode: Opposing effect of thermal annealing on Li and Na cyclability. *J. Phys. Chem. C* **2014**, *118* (49), 28401–28408.
27. Ding, J., Wang, H. L., Li, Z., Kohandehghan, A., Cui, K., Xu, Z. W., Zahiri, B. et al. Carbon nanosheet frameworks derived from peat moss as high performance sodium ion battery anodes. *ACS Nano* **2013**, *7* (12), 11004–11015.
28. Wang, Z. H., Qie, L., Yuan, L. X., Zhang, W. X., Hu, X. L., Huang, Y. H. Functionalized N-doped interconnected carbon nanofibers as an anode material for sodium-ion storage with excellent performance. *Carbon* **2013**, *55*, 328–334.
29. Xu, J. T., Wang, M., Wickramaratne, N. P., Jaroniec, M., Dou, S. X., Dai, L. M. High-performance sodium ion batteries based on a 3D anode from nitrogen-doped graphene foams. *Adv. Mater.* **2015**, *27* (12), 2042–2048.
30. Wang, S. Q., Xia, L., Yu, L., Zhang, L., Wang, H. H., Lou, X. W. Free-standing nitrogen-doped carbon nanofiber films: Integrated electrodes for sodium-ion batteries with ultralong cycle life and superior rate capability. *Adv. Energy Mater.* **2016**, *6* (7).
31. Li, W., Zhou, M., Li, H. M., Wang, K. L., Cheng, S. J., Jiang, K. A high performance sulfur-doped disordered carbon anode for sodium ion batteries. *Energ. Environ. Sci.* **2015**, *8* (10), 2916–2921.
32. Shen, F., Luo, W., Dai, J. Q., Yao, Y. G., Zhu, M. W., Hitz, E., Tang, Y. F., Chen, Y. F. et al. Ultra-thick, low-tortuosity, and mesoporous wood carbon anode for high-performance sodium-ion batteries. *Adv. Energy Mater.* **2016**, *6* (14).
33. Lotfabad, E. M., Ding, J., Cui, K., Kohandehghan, A., Kalisvaart, W. P., Hazelton, M., Mitlin, D. High-density sodium and lithium ion battery anodes from banana peels. *ACS Nano* **2014**, *8* (7), 7115–7129.
34. Li, H. B., Shen, F., Luo, W., Dai, J. Q., Han, X. G., Chen, Y. N., Yao, Y. G. et al. Carbonized-leaf membrane with anisotropic surfaces for sodium-ion battery. *ACS Appl. Mater. Inter.* **2016**, *8* (3), 2204–2210.

35. Wu, L. M., Buchholz, D., Vaalma, C., Giffin, G. A., Passerini, S. Apple-biowaste-derived hard carbon as a powerful anode material for Na-ion batteries. *Chemelectrochem* **2016**, *3* (2), 292–298.
36. Yang, T. Z., Qian, T., Wang, M. F., Shen, X. W., Xu, N., Sun, Z. Z., Yan, C. L. A sustainable route from biomass byproduct okara to high content nitrogen-doped carbon sheets for efficient sodium ion batteries. *Adv. Mater.* **2016**, *28* (3), 539.
37. Hong, K. L., Qie, L., Zeng, R., Yi, Z. Q., Zhang, W., Wang, D., Yin, W. et al. Biomass derived hard carbon used as a high performance anode material for sodium ion batteries. *J. Mater. Chem. A* **2014**, *2* (32), 12733–12738.
38. Senguttuvan, P., Rousse, G., Seznec, V., Tarascon, J. M., Palacin, M. R. $Na_2Ti_3O_7$: Lowest voltage ever reported oxide insertion electrode for sodium ion batteries. *Chem. Mater.* **2011**, *23* (18), 4109–4111.
39. Guo, S. H., Yi, J., Sun, Y., Zhou, H. S. Recent advances in titanium-based electrode materials for stationary sodium-ion batteries. *Energ. Environ. Sci.* **2016**, *9* (10), 2978–3006.
40. Xu, J., Ma, C. Z., Balasubramanian, M., Meng, Y. S. Understanding $Na_2Ti_3O_7$ as an ultra-low voltage anode material for a Na-ion battery. *Chem. Commun.* **2014**, *50* (83), 12564–12567.
41. Rudola, A., Saravanan, K., Mason, C. W., Balaya, P. $Na_2Ti_3O_7$: An intercalation based anode for sodium-ion battery applications. *J. Mater. Chem. A* **2013**, *1* (7), 2653–2662.
42. Rudola, A., Sharma, N., Balaya, P. Introducing a 0.2 V sodium-ion battery anode: The $Na_2Ti_3O_7$ to Na-3 (-) xTi_3O_7 pathway. *Electrochem. Commun.* **2015**, *61*, 10–13.
43. Pan, H. L., Lu, X., Yu, X. Q., Hu, Y. S., Li, H., Yang, X. Q., Chen, L. Q. Sodium storage and transport properties in layered $Na_2Ti_3O_7$ for room-temperature sodium-ion batteries. *Adv. Energy Mater.* **2013**, *3* (9), 1186–1194.
44. Munoz-Marquez, M. A., Zarrabeitia, M., Castillo-Martinez, E., Eguia-Barrio, A., Rojo, T., Casas-Cabanas, M. Composition and evolution of the solid-electrolyte interphase in $Na_2Ti_3O_7$ electrodes for Na-ion batteries: XPS and auger parameter analysis. *ACS Appl. Mater. Inter.* **2015**, *7* (14), 7801–7808.
45. Zou, W., Li, J. W., Deng, Q. J., Xue, J., Dai, X. Y., Zhou, A. J., Li, J. Z. Microspherical $Na_2Ti_3O_7$ prepared by spray-drying method as anode material for sodium-ion battery. *Solid State Ion* **2014**, *262*, 192–196.
46. Rudola, A., Saravanan, K., Devaraj, S., Gong, H., Balaya, P. $Na_2Ti_6O_{13}$: A potential anode for grid-storage sodium-ion batteries. *Chem. Commun.* **2013**, *49* (67), 7451–7453.
47. Shen, K., Wagemaker, M. $Na_{2+x}Ti_6O_{13}$ as potential negative electrode material for Na-ion batteries. *Inorg. Chem.* **2014**, *53* (16), 8250–8256.
48. Shirpour, M., Cabana, J., Doeff, M. New materials based on a layered sodium titanate for dual electrochemical Na and Li intercalation systems. *Energ. Environ. Sci.* **2013**, *6* (8), 2538–2547.
49. Woo, S. H., Park, Y., Choi, W. Y., Choi, N. S., Nam, S., Park, B., Lee, K. T. Trigonal $Na_4Ti_5O_{12}$ phase as an intercalation host for rechargeable batteries. *J. Electrochem. Soc.* **2012**, *159* (12), A2016–A2023.
50. Naeyaert, P. J. P., Avdeev, M., Sharma, N., Ben Yahia, H., Ling, C. D. Synthetic, structural, and electrochemical study of monoclinic $Na_4Ti_5O_{12}$ as a sodium-ion battery anode material. *Chem. Mater.* **2014**, *26* (24), 7067–7072.
51. Wu, D., Li, X., Xu, B., Twu, N., Liu, L., Ceder, G. $NaTiO_2$: A layered anode material for sodium-ion batteries. *Energ. Environ. Sci.* **2015**, *8* (1), 195–202.

52. Wang, Y. S., Yu, X. Q., Xu, S. Y., Bai, J. M., Xiao, R. J., Hu, Y. S., Li, H., Yang, X. Q., Chen, L. Q., Huang, X. J. A zero-strain layered metal oxide as the negative electrode for long-life sodium-ion batteries. *Nat. Commun.* **2013**, *4*, 2365.
53. Umebayashi, T., Yamaki, T., Itoh, H., Asai, K. Band gap narrowing of titanium dioxide by sulfur doping. *Appl. Phys. Lett.* **2002**, *81* (3), 454–456.
54. Jung, H. G., Myung, S. T., Yoon, C. S., Son, S. B., Oh, K. H., Amine, K., Scrosati, B., Sun, Y. K. Microscale spherical carbon-coated $Li_4Ti_5O_{12}$ as ultra high power anode material for lithium batteries. *Energ. Environ. Sci.* **2011**, *4* (4), 1345–1351.
55. Zhao, L., Pan, H. L., Hu, Y. S., Li, H., Chen, L. Q. Spinel lithium titanate ($Li_4Ti_5O_{12}$) as novel anode material for room-temperature sodium-ion battery. *Chinese Phys. B* **2012**, *21* (2).
56. Sun, Y., Zhao, L., Pan, H. L., Lu, X., Gu, L., Hu, Y. S., Li, H. et al. Direct atomic-scale confirmation of three-phase storage mechanism in $Li_4Ti_5O_{12}$ anodes for room-temperature sodium-ion batteries. *Nat. Commun.* **2013**, *4*.
57. Kitta, M., Kuratani, K., Tabuchi, M., Kataoka, R., Kiyobayashi, T., Kohyama, M. Design of a sodium-ion cell with a carbon-free $Li_4Ti_5O_{12}$ negative electrode. *Electrochemistry* **2015**, *83* (11), 989–992.
58. Kim, K. T., Yu, C. Y., Yoon, C. S., Kim, S. J., Sun, Y. K., Myung, S. T. Carbon-coated $Li_4Ti_5O_{12}$ nanowires showing high rate capability as an anode material for rechargeable sodium batteries. *Nano Energ.* **2015**, *12*, 725–734.
59. Yu, X. Q., Pan, H. L., Wan, W., Ma, C., Bai, J. M., Meng, Q. P., Ehrlich, S. N., Hu, Y. S., Yang, X. Q. A size-dependent sodium storage mechanism in $L(i)_4Ti(_5)O(_{12})$ investigated by, a novel characterization technique combining in situ X-ray diffraction and chemical sodiation. *Nano Lett.* **2013**, *13* (10), 4721–4727.
60. Hasegawa, G., Kanamori, K., Kiyomura, T., Kurata, H., Nakanishi, K., Abe, T. Hierarchically porous $Li_4Ti_5O_{12}$ anode materials for Li- and Na-ion batteries: Effects of nanoarchitectural design and temperature dependence of the rate capability. *Adv. Energ. Mater.* **2015**, *5* (1).
61. Yu, P. F., Li, C. L., Guo, X. X. Sodium storage and pseudocapacitive charge in textured $Li_4Ti_5O_{12}$ thin films. *J Phys. Chem. C* **2014**, *118* (20), 10616–10624.
62. Chen, C. J., Xu, H. H., Zhou, T. F., Guo, Z. P., Chen, L. N., Yan, M. Y., Mai, L. Q. et al. Integrated intercalation-based and interfacial sodium storage in graphene-wrapped porous $Li_4Ti_5O_{12}$ nanofibers composite aerogel. *Adv. Energy Mater.* **2016**, *6* (13).
63. Zhou, Q., Liu, L., Tan, J. L., Yan, Z. C., Huang, Z. F., Wang, X. Y. Synthesis of lithium titanate nanorods as anode materials for lithium and sodium ion batteries with superior electrochemical performance. *J. Power Sources* **2015**, *283*, 243–250.
64. Yang, L. Y., Li, H. Z., Liu, J., Tang, S. S., Lu, Y. K., Li, S. T., Min, J., Yan, N., Lei, M. $Li_4Ti_5O_{12}$ nanosheets as high-rate and long-life anode materials for sodium-ion batteries. *J. Mater. Chem. A* **2015**, *3* (48), 24446–24452.
65. Feng, X. Y., Zou, H. L., Xiang, H. F., Guo, X., Zhou, T. P., Wu, Y. C., Xu, W. et al. Ultrathin $Li_4Ti_5O_{12}$ nanosheets as anode materials for lithium and sodium storage. *ACS Appl. Mater. Inter.* **2016**, *8* (26), 16718–16726.
66. Legrain, F., Malyi, O., Manzhos, S. Insertion energetics of lithium, sodium, and magnesium in crystalline and amorphous titanium dioxide: A comparative first-principles study. *J. Power Sources* **2015**, *278*, 197–202.
67. Su, D. W., Dou, S. X., Wang, G. X. Anatase TiO_2: Better anode material than amorphous and rutile phases of TiO_2 for Na-ion batteries. *Chem. Mater.* **2015**, *27* (17), 6022–6029.

68. Lunell, S., Stashans, A., Ojamae, L., Lindstrom, H., Hagfeldt, A. Li and Na diffusion in TiO$_2$ from quantum chemical theory versus electrochemical experiment. *J. Am. Chem. Soc.* **1997**, *119* (31), 7374–7380.
69. Xu, Y., Lotfabad, E. M., Wang, H. L., Farbod, B., Xu, Z. W., Kohandehghan, A., Mitlin, D. Nanocrystalline anatase TiO$_2$: A new anode material for rechargeable sodium ion batteries. *Chem. Commun.* **2013**, *49* (79), 8973–8975.
70. Xiong, H., Slater, M. D., Balasubramanian, M., Johnson, C. S., Rajh, T. Amorphous TiO$_2$ nanotube anode for rechargeable sodium ion batteries. *J. Phys. Chem. Lett.* **2011**, *2* (20), 2560–2565.
71. Wu, L. M., Buchholz, D., Bresser, D., Chagas, L. G., Passerini, S. Anatase TiO$_2$ nanoparticles for high power sodium-ion anodes. *J. Power Sources* **2014**, *251*, 379–385.
72. Kim, K. T., Ali, G., Chung, K. Y., Yoon, C. S., Yashiro, H., Sun, Y. K., Lu, J., Amine, K., Myung, S. T. Anatase titania nanorods as an intercalation anode material for rechargeable sodium batteries. *Nano Lett.* **2014**, *14* (2), 416–422.
73. Gonzalez, J. R., Alcantara, R., Nacimiento, F., Ortiz, G. F., Tirado, J. L. Microstructure of the epitaxial film of anatase nanotubes obtained at high voltage and the mechanism of its electrochemical reaction with sodium. *Crystengcomm* **2014**, *16* (21), 4602–4609.
74. Wu, L. M., Bresser, D., Buchholz, D., Giffin, G. A., Castro, C. R., Ochel, A., Passerini, S. Unfolding the mechanism of sodium insertion in anatase TiO$_2$ nanoparticles. *Adv. Energy Mater.* **2015**, *5* (2).
75. Yeo, Y., Jung, J. W., Park, K., Kim, I. D. Graphene-wrapped anatase TiO$_2$ nanofibers as high-rate and long-cycle-life anode material for sodium ion batteries. *Sci. Rep.-UK* **2015**, *5*.
76. Wang, B. F., Zhao, F., Du, G. D., Porter, S., Liu, Y., Zhang, P., Cheng, Z. X., Liu, H. K., Huang, Z. G. Boron-doped anatase TiO$_2$ as a high-performance anode material for sodium-ion batteries. *ACS Appl. Mater. Inter.* **2016**, *8* (25), 16009–16015.
77. Gopal, N. O., Lo, H. H., Ke, S. C. Chemical state and environment of boron dopant in B, N-codoped anatase TiO$_2$ nanoparticles: An avenue for probing diamagnetic dopants in TiO$_2$ by electron paramagnetic resonance spectroscopy. *J. Am. Chem. Soc.* **2008**, *130* (9), 2760.
78. Xu, Y., Zhou, M., Wen, L. Y., Wang, C. L., Zhao, H. P., Mi, Y., Liang, L. Y., Fu, Q., Wu, M. H., Lei, Y. Highly ordered three-dimensional Ni-TiO$_2$ nanoarrays as sodium ion battery anodes. *Chem. Mater.* **2015**, *27* (12), 4274–4280.
79. Yang, Y. C., Ji, X. B., Jing, M. J., Hou, H. S., Zhu, Y. R., Fang, L. B., Yang, X. M., Chen, Q. Y., Banks, C. E. Carbon dots supported upon N-doped TiO$_2$ nanorods applied into sodium and lithium ion batteries. *J. Mater. Chem. A* **2015**, *3* (10), 5648–5655.
80. Oh, S. M., Hwang, J. Y., Yoon, C. S., Lu, J., Amine, K., Belharouak, I., Sun, Y. K. High electrochemical performances of microsphere C-TiO$_2$ anode for sodium-ion battery. *ACS Appl. Mater. Inter.* **2014**, *6* (14), 11295–11301.
81. Yang, X. M., Wang, C., Yang, Y. C., Zhang, Y., Jia, X. N., Chen, J., Ji, X. B. Anatase TiO$_2$ nanocubes for fast and durable sodium ion battery anodes. *J. Mater. Chem. A* **2015**, *3* (16), 8800–8807.
82. Hong, Z. S., Zhou, K. Q., Huang, Z. G., Wei, M. D. Iso-oriented anatase TiO$_2$ mesocages as a high performance anode material for sodium-ion storage. *Sci. Rep.-UK* **2015**, *5*.

83. Tahir, M. N., Oschmann, B., Buchholz, D., Dou, X. W., Lieberwirth, I., Panthofer, M., Tremel, W., Zentel, R., Passerini, S. Extraordinary performance of carbon-coated anatase TiO$_2$ as sodium-ion anode. *Adv. Energy Mater.* **2016**, *6* (4).
84. Ge, Y. Q., Jiang, H., Zhu, J. D., Lu, Y., Chen, C., Hu, Y., Qiu, Y. P., Zhang, X. W. High cyclability of carbon-coated TiO$_2$ nanoparticles as anode for sodium-ion batteries. *Electrochim. Acta* **2015**, *157*, 142–148.
85. Ni, J. F., Fu, S. D., Wu, C., Maier, J., Yu, Y., Li, L. Self-supported nanotube arrays of sulfur-doped TiO$_2$ enabling ultrastable and robust sodium storage. *Adv. Mater.* **2016**, *28* (11), 2259–2265.
86. Hong, Z. S., Zhou, K. D., Zhang, J. W., Huang, Z. G., Wei, M. D. Facile synthesis of rutile TiO$_2$ mesocrystals with enhanced sodium storage properties. *J. Mater. Chem. A* **2015**, *3* (33), 17412–17416.
87. Hong, Z. S., Hong, J. X., Xie, C. B., Huang, Z. G., Wei, M. D. Hierarchical rutile TiO$_2$ with mesocrystalline structure for Li-ion and Na-ion storage. *Electrochim. Acta* **2016**, *202*, 203–208.
88. Usui, H., Yoshioka, S., Wasada, K., Shimizu, M., Sakaguchi, H. Nb-doped rutile TiO$_2$: A potential anode material for Na-ion battery. *ACS Appl. Mater. Inter.* **2015**, *7* (12), 6567–6573.
89. Zhang, Y., Pu, X. L., Yang, Y. C., Zhu, Y. R., Hou, H. R., Jing, M. J., Yang, X. M., Chena, J., Ji, X. B. An electrochemical investigation of rutile TiO$_2$ microspheres anchored by nanoneedle clusters for sodium storage. *Phys. Chem. Chem. Phys.* **2015**, *17* (24), 15764–15770.
90. Huang, J. P., Yuan, D. D., Zhang, H. Z., Cao, Y. L., Li, G. R., Yang, H. X., Gao, X. P. Electrochemical sodium storage of TiO$_2$(B) nanotubes for sodium ion batteries. *RSC Adv.* **2013**, *3* (31), 12593–12597.
91. Chen, C. J., Wen, Y. W., Hu, X. L., Ji, X. L., Yan, M. Y., Mai, L. Q., Hu, P., Shan, B., Huang, Y. H. Na+ intercalation pseudocapacitance in graphene-coupled titanium oxide enabling ultra-fast sodium storage and long-term cycling. *Nat. Commun.* **2015**, *6*.
92. Wu, L. M., Bresser, D., Buchholz, D., Passerini, S. Nanocrystalline TiO$_2$(B) as anode material for sodium-ion batteries. *J. Electrochem. Soc.* **2015**, *162* (2), A3052–A3058.
93. Zhang, Y., Foster, C. W., Banks, C. E., Shao, L. D., Hou, H. S., Zou, G. Q., Chen, J., Huang, Z. D., Ji, X. B. Graphene-rich wrapped petal-like rutile TiO$_2$ tuned by carbon dots for high-performance sodium storage. *Adv. Mater.* **2016**, *28* (42), 9391.
94. Sondergaard, M., Dalgaard, K. J., Bojesen, E. D., Wonsyld, K., Dahl, S., Iversen, B. B. In situ monitoring of TiO$_2$(B)/anatase nanoparticle formation and application in Li-ion and Na-ion batteries. *J. Mater. Chem. A* **2015**, *3* (36), 18667–18674.
95. Hwang, J. Y., Myung, S. T., Lee, J. H., Abouimrane, A., Belharouak, I., Sun, Y. K. Ultrafast sodium storage in anatase TiO$_2$ nanoparticles embedded on carbon nanotubes. *Nano Energy* **2015**, *16*, 218–226.
96. Zhao, L., Pan, H. L., Hu, Y. S., Li, H., Chen, L. Q. Spinel lithium titanate (Li$_4$Ti$_5$O$_{12}$) as novel anode material for room-temperature sodium-ion battery (vol 21, 028201, 2012). *Chinese Phys. B* **2012**, *21* (7).
97. Feng, L. L., Li, G. D., Liu, Y. P., Wu, Y. Y., Chen, H., Wang, Y., Zou, Y. C., Wang, D. J., Zou, X. X. Carbon-armored Co9S8 nanoparticles as all-pH efficient and durable H-2-evolving electrocatalysts. *ACS Appl. Mater. Inter.* **2015**, *7* (1), 980–988.
98. He, D., Wu, D. N., Gao, J., Wu, X. M., Zeng, X. Q., Ding, W. J. Flower-like CoS with nanostructures as a new cathode-active material for rechargeable magnesium batteries. *J. Power Sources* **2015**, *294*, 643–649.

99. Liu, J., Wu, C., Xiao, D. D., Kopold, P., Gu, L., van Aken, P. A., Maier, J., Yu, Y. MOF-derived hollow Co9S8 nanoparticles embedded in graphitic carbon nanocages with superior Li-ion storage. *Small* **2016**, *12* (17), 2354–2364.
100. Shadike, Z., Cao, M. H., Ding, F., Sang, L., Fu, Z. W. Improved electrochemical performance of CoS$_2$-MWCNT nanocomposites for sodium-ion batteries. *Chem. Commun.* **2015**, *51* (52), 10486–10489.
101. Yin, D. M., Huang, G., Zhang, F. F., Qin, Y. L., Na, Z. L., Wu, Y. M., Wang, L. M. Coated/sandwiched rGO/CoSx composites derived from metal-organic frameworks/GO as advanced anode materials for lithium-ion batteries. *Chem.-Eur. J.* **2016**, *22* (4), 1467–1474.
102. Zhou, Q., Liu, L., Guo, G. X., Yan, Z. C., Tan, J. L., Huang, Z. F., Chen, X. Y., Wang, X. Y. Sandwich-like cobalt sulfide-graphene composite—An anode material with excellent electrochemical performance for sodium ion batteries. *RSC Adv.* **2015**, *5* (88), 71644–71651.
103. Zhou, Q., Liu, L., Huang, Z. F., Yi, L. G., Wang, X. Y., Cao, G. Z. Co$_3$S$_4$@polyaniline nanotubes as high-performance anode materials for sodium ion batteries. *J. Mater. Chem. A* **2016**, *4* (15), 5505–5516.
104. Park, J., Kim, J. S., Park, J. W., Nam, T. H., Kim, K. W., Ahn, J. H., Wang, G., Ahn, H. J. Discharge mechanism of MoS$_2$ for sodium ion battery: Electrochemical measurements and characterization. *Electrochim. Acta* **2013**, *92*, 427–432.
105. Hu, Z., Wang, L. X., Zhang, K., Wang, J. B., Cheng, F. Y., Tao, Z. L., Chen, J. MoS$_2$ nanoflowers with expanded interlayers as high-performance anodes for sodium-ion batteries. *Angew. Chem. Int. Edit.* **2014**, *53* (47), 12794–12798.
106. Ryu, W. H., Jung, J. W., Park, K., Kim, S. J., Kim, I. D., Vine-like MoS$_2$ anode materials self-assembled from 1-D nanofibers for high capacity sodium rechargeable batteries. *Nanoscale* **2014**, *6* (19), 10975–10981.
107. Wang, Y. X., Seng, K. H., Chou, S. L., Wang, J. Z., Guo, Z. P., Wexler, D., Liu, H. K., Dou, S. X. Reversible sodium storage via conversion reaction of a MoS$_2$-C composite. *Chem. Commun.* **2014**, *50* (73), 10730–10733.
108. Wang, Y. X., Chou, S. L., Wexler, D., Liu, H. K., Dou, S. X. High-performance sodium-ion batteries and sodium-ion pseudocapacitors based on MoS$_2$/graphene composites. *Chem.-Eur. J.* **2014**, *20* (31), 9607–9612.
109. Zhang, S., Yu, X. B., Yu, H. L., Chen, Y. J., Gao, P., Li, C. Y., Zhu, C. L. Growth of ultrathin MoS$_2$ nanosheets with expanded spacing of (002) plane on carbon nanotubes for high-performance sodium-ion battery anodes. *ACS Appl. Mater. Inter.* **2014**, *6* (24), 21880–21885.
110. Xie, X. Q., Ao, Z. M., Su, D. W., Zhang, J. Q., Wang, G. X. MoS$_2$/graphene composite anodes with enhanced performance for sodium-ion batteries: The role of the two-dimensional heterointerface. *Adv. Funct. Mater.* **2015**, *25* (9), 1393–1403.
111. Su, D. W., Dou, S. X., Wang, G. X. Ultrathin MoS$_2$ nanosheets as anode materials for sodium-ion batteries with superior performance. *Adv. Energy Mater.* **2015**, *5* (6).
112. Kalluri, S., Seng, K. H., Guo, Z. P., Du, A., Konstantinov, K., Liu, H. K., Dou, S. X. Sodium and lithium storage properties of spray-dried molybdenum disulfide-graphene hierarchical microspheres. *Sci. Rep.-UK* **2015**, *5*.
113. Choi, S. H., Ko, Y. N., Lee, J. K., Kang, Y. C. 3D MoS$_2$-graphene microspheres consisting of multiple nanospheres with superior sodium ion storage properties. *Adv. Funct. Mater.* **2015**, *25* (12), 1780–1788.

114. Kim, T. B., Jung, W. H., Ryu, H. S., Kim, K. W., Ahn, J. H., Cho, K. K., Cho, G. B., Nam, T. H., Ahn, I. S., Ahn, H. J. Electrochemical characteristics of Na/FeS$_2$ battery by mechanical alloying. *J. Alloy Compd.* **2008**, *449* (1–2), 304–307.
115. Kitajou, A., Yamaguchi, J., Hara, S., Okada, S. Discharge/charge reaction mechanism of a pyrite-type FeS$_2$ cathode for sodium secondary batteries. *J. Power Sources* **2014**, *247*, 391–395.
116. Zhu, Y. J., Suo, L. M., Gao, T., Fan, X. L., Han, F. D., Wang, C. S. Ether-based electrolyte enabled Na/FeS$_2$ rechargeable batteries. *Electrochem. Commun.* **2015**, *54*, 18–22.
117. Hu, Z., Zhu, Z. Q., Cheng, F. Y., Zhang, K., Wang, J. B., Chen, C. C., Chen, J. Pyrite FeS$_2$ for high-rate and long-life rechargeable sodium batteries. *Energ. Environ. Sci.* **2015**, *8* (4), 1309–1316.
118. Wei, X., Li, W. H., Shi, J. A., Gu, L., Yu, Y. FeS@C on carbon cloth as flexible electrode for both lithium and sodium storage. *ACS Appl. Mater. Inter.* **2015**, *7* (50), 27804–27809.
119. Wu, L., Lu, H. Y., Xiao, L. F., Qian, J. F., Ai, X. P., Yang, H. X., Cao, Y. L. A tin(II) sulfide-carbon anode material based on combined conversion and alloying reactions for sodium-ion batteries. *J. Mater. Chem. A* **2014**, *2* (39), 16424–16428.
120. Lu, Y. C., Ma, C., Alvarado, J., Dimov, N., Meng, Y. S., Okada, S. Improved electrochemical performance of tin-sulfide anodes for sodium-ion batteries. *J Mater. Chem. A* **2015**, *3* (33), 16971–16977.
121. Wu, L., Hu, X. H., Qian, J. F., Pei, F., Wu, F. Y., Mao, R. J., Ai, X. P., Yang, H. X., Cao, Y. L. A Sn-SnS-C nanocomposite as anode host materials for Na-ion batteries. *J. Mater. Chem. A* **2013**, *1* (24), 7181–7184.
122. Qu, B. H., Ma, C. Z., Ji, G., Xu, C. H., Xu, J., Meng, Y. S., Wang, T. H., Lee, J. Y. Layered SnS$_2$-reduced graphene oxide composite—A high-capacity, high-rate, and long-cycle life sodium-ion battery anode material. *Adv. Mater.* **2014**, *26* (23), 3854–3859.
123. Wang, J. J., Luo, C., Mao, J. F., Zhu, Y. J., Fan, X. L., Gao, T., Mignerey, A. C., Wang, C. S. Solid-state fabrication of SnS$_2$/C nanospheres for high-performance sodium ion battery anode. *ACS Appl. Mater. Inter.* **2015**, *7* (21), 11476–11481.
124. Prikhodchenko, P. V., Yu, D. Y. W., Batabyal, S. K., Uvarov, V., Gun, J., Sladkevich, S., Mikhaylov, A. A., Medvedev, A. G., Lev, O. Nanocrystalline tin disulfide coating of reduced graphene oxide produced by the peroxostannate deposition route for sodium ion battery anodes. *J. Mater. Chem. A* **2014**, *2* (22), 8431–8437.
125. Pan, Q., Xie, J., Zhu, T. J., Cao, G. S., Zhao, X. B., Zhang, S. C. Reduced graphene oxide-induced recrystallization of NiS nanorods to nanosheets and the improved Na-storage properties. *Inorg. Chem.* **2014**, *53* (7), 3511–3518.
126. Xu, X. J., Ji, S. M., Gu, M. Z., Liu, J. In situ synthesis of MnS hollow microspheres on reduced graphene oxide sheets as high-capacity and long-life anodes for Li- and Na-ion batteries. *ACS Appl. Mater. Inter.* **2015**, *7* (37), 20957–20964.
127. Wang, T. S., Hu, P., Zhang, C. J., Du, H. P., Zhang, Z. H., Wang, X. G., Chen, S. G., Xiong, J. W., Cui, G. L. Nickel disulfide-graphene nanosheets composites with improved electrochemical performance for sodium ion battery. *ACS Appl. Mater. Inter.* **2016**, *8* (12), 7811–7817.
128. Su, D. W., Dou, S. X., Wang, G. X. WS$_2$@graphene nanocomposites as anode materials for Na-ion batteries with enhanced electrochemical performances. *Chem. Commun.* **2014**, *50* (32), 4192–4195.

129. Su, D. W., Kretschmer, K., Wang, G. X. Improved electrochemical performance of Na-ion batteries in ether-based electrolytes: A case study of ZnS nanospheres. *Adv. Energy Mater.* **2016**, *6* (2).
130. Lu, Y. Y., Zhao, Q., Zhang, N., Lei, K. X., Li, F. J., Chen, J. Facile spraying synthesis and high-performance sodium storage of mesoporous MoS_2/C microspheres. *Adv. Funct. Mater.* **2016**, *26* (6), 911–918.
131. Kim, T. B., Choi, J. W., Ryu, H. S., Cho, G. B., Kim, K. W., Ahn, J. H., Cho, K. K., Ahn, H. J. Electrochemical properties of sodium/pyrite battery at room temperature. *J. Power Sources* **2007**, *174* (2), 1275–1278.
132. Douglas, A., Carter, R., Oakes, L., Share, K., Cohn, A. P., Pint, C. L. Ultrafine iron pyrite (FeS_2) nanocrystals improve sodium-sulfur and lithium-sulfur conversion reactions for efficient batteries. *ACS Nano* **2015**, *9* (11), 11156–11165.
133. Lee, S. Y., Kang, Y. C. Sodium-ion storage properties of FeS-reduced graphene oxide composite powder with a crumpled structure. *Chem.-Eur. J.* **2016**, *22* (8), 2769–2774.
134. Walter, M., Zund, T., Kovalenko, M. V. Pyrite (FeS_2) nanocrystals as inexpensive high-performance lithium-ion cathode and sodium-ion anode materials. *Nanoscale* **2015**, *7* (20), 9158–9163.
135. Wu, L., Lu, H. Y., Xiao, L. F., Ai, X. P., Yang, H. X., Cao, Y. L. Improved sodium-storage performance of stannous sulfide@reduced graphene oxide composite as high capacity anodes for sodium-ion batteries. *J. Power Sources* **2015**, *293*, 784–789.
136. Zhu, C. B., Kopold, P., Li, W. H., van Aken, P. A., Maier, J., Yu, Y. A general strategy to fabricate carbon-coated 3D porous interconnected metal sulfides: Case study of SnS/C nanocomposite for high-performance lithium and sodium ion batteries. *Adv. Sci.* **2015**, *2* (12).
137. Xie, X. Q., Su, D. W., Chen, S. Q., Zhang, J. Q., Dou, S. X., Wang, G. X. SnS_2 Nanoplatelet@Graphene nanocomposites as high-capacity anode materials for sodium-ion batteries. *Chem.-Asian J.* **2014**, *9* (6), 1611–1617.
138. Zhang, Y. D., Zhu, P. Y., Huang, L. L., Xie, J., Zhang, S. C., Cao, G. S., Zhao, X. B. Few-layered SnS_2 on few-layered reduced graphene oxide as Na-ion battery anode with ultralong cycle life and superior rate capability. *Adv. Funct. Mater.* **2015**, *25* (3), 481–489.
139. Liu, Y. C., Kang, H. Y., Jiao, L. F., Chen, C. C., Cao, K. Z., Wang, Y. J., Yuan, H. T. Exfoliated-SnS_2 restacked on graphene as a high-capacity, high-rate, and long-cycle life anode for sodium ion batteries. *Nanoscale* **2015**, *7* (4), 1325–1332.
140. Kim, J. S., Kim, D. Y., Cho, G. B., Nam, T. H., Kim, K. W., Ryu, H. S., Ahn, J. H., Ahn, H. J. The electrochemical properties of copper sulfide as cathode material for rechargeable sodium cell at room temperature. *J. Power Sources* **2009**, *189* (1), 864–868.
141. Alcantara, R., Jaraba, M., Lavela, P., Tirado, J. L. $NiCo_2O_4$ spinel: First report on a transition metal oxide for the negative electrode of sodium-ion batteries. *Chem. Mater.* **2002**, *14* (7), 2847.
142. Hariharan, S., Saravanan, K., Ramar, V., Balaya, P. A rationally designed dual role anode material for lithium-ion and sodium-ion batteries: Case study of eco-friendly Fe_3O_4. *Phys. Chem. Chem. Phys.* **2013**, *15* (8), 2945–2953.
143. Liu, S. H., Wang, Y. W., Dong, Y. F., Zhao, Z. B., Wang, Z. Y., Qiu, J. S. Ultrafine Fe_3O_4 quantum dots on hybrid carbon nanosheets for long-life, high-rate alkali-metal storage. *Chemelectrochem* **2016**, *3* (1), 38–44.

144. Kumar, P. R., Jung, Y. H., Bharathi, K. K., Lim, C. H., Kim, D. K. High capacity and low cost spinel Fe_3O_4 for the Na-ion battery negative electrode materials. *Electrochim. Acta* **2014**, *146*, 503–510.
145. Jian, Z. L., Zhao, B., Liu, P., Li, F. J., Zheng, M. B., Chen, M. W., Shi, Y., Zhou, H. S. Fe_2O_3 nanocrystals anchored onto graphene nanosheets as the anode material for low-cost sodium-ion batteries. *Chem. Commun.* **2014**, *50* (10), 1215–1217.
146. Liu, X. J., Chen, T. Q., Chu, H. P., Niu, L. Y., Sun, Z., Pan, L. K., Sun, C. Q. Fe_2O_3-reduced graphene oxide composites synthesized via microwave-assisted method for sodium ion batteries. *Electrochim. Acta* **2015**, *166*, 12–16.
147. Philippe, B., Valvo, M., Lindgren, F., Rensmo, H., Edstrom, K. Investigation of the electrode/electrolyte interface of Fe_2O_3 composite electrodes: Li versus Na batteries. *Chem. Mater.* **2014**, *26* (17), 5028–5041.
148. Rahman, M. M., Sultana, I., Chen, Z. Q., Srikanth, M., Li, L. H., Dai, X. J. J., Chen, Y. *Ex situ* electrochemical sodiation/desodiation observation of Co_3O_4 anchored carbon nanotubes: A high performance sodium-ion battery anode produced by pulsed plasma in a liquid. *Nanoscale* **2015**, *7* (30), 13088–13095.
149. Jian, Z. L., Liu, P., Li, F. J., Chen, M. W., Zhou, H. S. Monodispersed hierarchical Co_3O_4 spheres intertwined with carbon nanotubes for use as anode materials in sodium-ion batteries. *J. Mater. Chem. A* **2014**, *2* (34), 13805–13809.
150. Gu, M., Kushima, A., Shao, Y. Y., Zhang, J. G., Liu, J., Browning, N. D., Li, J., Wang, C. M. Probing the failure mechanism of SnO_2 nanowires for sodium-ion batteries. *Nano Lett.* **2013**, *13* (11), 5203–5211.
151. Dirican, M., Lu, Y., Ge, Y. Q., Yildiz, O., Zhang, X. W. Carbon-confined Sno(2)-electrodeposited porous carbon nanofiber composite as high-capacity sodium-ion battery anode material. *ACS Appl. Mater. Inter.* **2015**, *7* (33), 18387–18396.
152. Su, D. W., Xie, X. Q., Wang, G. X. Hierarchical mesoporous SnO microspheres as high capacity anode materials for sodium-ion batteries. *Chem.-Eur. J.* **2014**, *20* (11), 3192–3197.
153. Kalubarme, R. S., Lee, J. Y., Park, C. J., Carbon encapsulated tin oxide nanocomposites: An efficient anode for high performance sodium-ion batteries. *ACS Appl. Mater. Inter.* **2015**, *7* (31), 17226–17237.
154. Wang, L. J., Zhang, K., Hu, Z., Duan, W. C., Cheng, F. Y., Chen, J. Porous CuO nanowires as the anode of rechargeable Na-ion batteries. *Nano Res.* **2014**, *7* (2), 199–208.
155. Kang, W. P., Zhang, Y., Fan, L. L., Zhang, L. L., Dai, F. N., Wang, R. M., Sun, D. F. Metal organic framework derived porous hollow Co_3O_4/N-C polyhedron composite with excellent energy storage capability. *ACS Appl. Mater. Inter.* **2017**, *9* (12), 10602–10609.
156. Lu, Y. Y., Zhang, N., Zhao, Q., Liang, J., Chen, J. Micro-nanostructured CuO/C spheres as high-performance anode materials for Na-ion batteries. *Nanoscale* **2015**, *7* (6), 2770–2776.
157. Zou, F., Chen, Y. M., Liu, K. W., Yu, Z. T., Liang, W. F., Bhaway, S. M., Gao, M., Zhu, Y. Metal organic frameworks derived hierarchical hollow NiO/Ni/graphene composites for lithium and sodium storage. *ACS Nano* **2016**, *10* (1), 377–386.
158. Komaba, S., Mikumo, T., Yabuuchi, N., Ogata, A., Yoshida, H., Yamada, Y. Electrochemical insertion of Li and Na ions into nanocrystalline Fe_3O_4 and alpha-Fe_2O_3 for rechargeable batteries. *J. Electrochem. Soc.* **2010**, *157* (1), A60–A65.

159. Park, D. Y., Myung, S. T. Carbon-coated magnetite embedded on carbon nanotubes for rechargeable lithium and sodium batteries. *ACS Appl. Mater. Inter.* **2014**, *6* (14), 11749–11757.
160. Koo, B., Chattopadhyay, S., Shibata, T., Prakapenka, V. B., Johnson, C. S., Rajh, T., Shevchenko, E. V. Intercalation of sodium ions into hollow iron oxide nanoparticles. *Chem. Mater.* **2013**, *25* (2), 245–252.
161. Ming, J., Ming, H., Yang, W. J., Kwak, W. J., Park, J. B., Zheng, J. W., Sun, Y. K. A sustainable iron-based sodium ion battery of porous carbon-Fe$_3$O$_4$/Na$_2$FeP$_2$O$_7$ with high performance. *RSC Adv.* **2015**, *5* (12), 8793–8800.
162. Huang, B., Tai, K. P., Zhang, M. G., Xiao, Y. R., Dillon, S. J. Comparative study of Li and Na electrochemical reactions with iron oxide nanowires. *Electrochim. Acta* **2014**, *118*, 143–149.
163. Valvo, M., Lindgren, F., Lafont, U., Bjorefors, F., Edstrom, K. Towards more sustainable negative electrodes in Na-ion batteries via nanostructured iron oxide. *J. Power Sources* **2014**, *245*, 967–978.
164. Rahman, M. M., Glushenkov, A. M., Ramireddy, T., Chen, Y. Electrochemical investigation of sodium reactivity with nanostructured Co$_3$O$_4$ for sodium-ion batteries. *Chem. Commun.* **2014**, *50* (39), 5057–5060.
165. Deng, Q. J., Wang, L. P., Li, J. Z. Electrochemical characterization of Co$_3$O$_4$/MCNTs composite anode materials for sodium-ion batteries. *J. Mater. Sci.* **2015**, *50* (11), 4142–4148.
166. Klavetter, K. C., Garcia, S., Dahal, N., Snider, J. L., de Souza, J. P., Cell, T. H., Cassara, M. A., Heller, A., Humphrey, S. M., Mullins, C. B. Li- and Na-reduction products of meso-Co$_3$O$_4$ form high-rate, stably cycling battery anode materials. *J. Mater. Chem. A* **2014**, *2* (34), 14209–14221.
167. Wang, Y. X., Lim, Y. G., Park, M. S., Chou, S. L., Kim, J. H., Liu, H. K., Dou, S. X., Kim, Y. J. Ultrafine SnO$_2$ nanoparticle loading onto reduced graphene oxide as anodes for sodium-ion batteries with superior rate and cycling performances. *J. Mater. Chem. A* **2014**, *2* (2), 529–534.
168. Lu, Y. C., Ma, C. Z., Alvarado, J., Kidera, T., Dimov, N., Meng, Y. S., Okada, S. Electrochemical properties of tin oxide anodes for sodium-ion batteries. *J. Power Sources* **2015**, *284*, 287–295.
169. Xie, X. Q., Chen, S. Q., Sun, B., Wang, C. Y., Wang, G. X. 3D networked tin oxide/graphene aerogel with a hierarchically porous architecture for high-rate performance sodium-ion batteries. *Chemsuschem* **2015**, *8* (17), 2948–2955.
170. Xie, X. Q., Su, D. W., Zhang, J. Q., Chen, S. Q., Mondal, A. K., Wang, G. X. A comparative investigation on the effects of nitrogen-doping into graphene on enhancing the electrochemical performance of SnO$_2$/graphene for sodium-ion batteries. *Nanoscale* **2015**, *7* (7), 3164–3172.
171. Yuan, S., Huang, X. L., Ma, D. L., Wang, H. G., Meng, F. Z., Zhang, X. B. Engraving copper foil to give large-scale binder-free porous CuO arrays for a high-performance sodium-ion battery anode. *Adv. Mater.* **2014**, *26* (14), 2273–2279.
172. Liu, H. H., Cao, F., Zheng, H., Sheng, H. P., Li, L., Wu, S. J., Liu, C., Wang, J. B. In situ observation of the sodiation process in CuO nanowires. *Chem. Commun.* **2015**, *51* (52), 10443–10446.
173. Zhang, Z., Feng, J. K., Ci, L. J., Tian, Y., Xiong, S. L. Mental-organic framework derived CuO hollow spheres as high performance anodes for sodium ion battery. *Mater Technol.* **2016**, *31* (9), 497–500.

174. Jiang, Y. Z., Hu, M. J., Zhang, D., Yuan, T. Z., Sun, W. P., Xu, B., Yan, M. Transition metal oxides for high performance sodium ion battery anodes. *Nano Energy* **2014**, *5*, 60–66.
175. Fullenwarth, J., Darwiche, A., Soares, A., Donnadieu, B., Monconduit, L. NiP3: A promising negative electrode for Li- and Na-ion batteries. *J. Mater. Chem. A* **2014**, *2* (7), 2050–2059.
176. Fan, M. P., Chen, Y., Xie, Y. H., Yang, T. Z., Shen, X. W., Xu, N., Yu, H. Y., Yan, C. L. Half-cell and full-cell applications of highly stable and binder-free sodium ion batteries based on Cu$_3$P nanowire anodes. *Adv. Funct. Mater.* **2016**, *26* (28), 5019–5027.
177. Qian, J. F., Xiong, Y., Cao, Y. L., Ai, X. P., Yang, H. X. Synergistic Na-storage reactions in Sn$_4$P$_3$ as a high-capacity, cycle-stable anode of Na-ion batteries. *Nano Lett.* **2014**, *14* (4), 1865–1869.
178. Li, W. J., Chou, S. L., Wang, J. Z., Kim, J. H., Liu, H. K., Dou, S. X. Sn$_4$+xP$_3$ @ Amorphous Sn-P composites as anodes for sodium-ion batteries with low cost, high capacity, long life, and superior rate capability. *Adv. Mater.* **2014**, *26* (24), 4037–4042.
179. Liu, J., Kopold, P., Wu, C., van Aken, P. A., Maier, J., Yu, Y. Uniform yolk-shell Sn$_4$P$_3$@C nanospheres as high-capacity and cycle-stable anode materials for sodium-ion batteries. *Energ. Environ. Sci.* **2015**, *8* (12), 3531–3538.
180. Mao, J. F., Fan, X. L., Luo, C., Wang, C. S. Building self-healing alloy architecture for stable sodium-ion battery anodes: A case study of tin anode materials. *ACS Appl. Mater. Inter.* **2016**, *8* (11), 7147–7155.
181. Kim, S. O., Manthiram, A. The facile synthesis and enhanced sodium-storage performance of a chemically bonded CuP$_2$/C hybrid anode. *Chem. Commun.* **2016**, *52* (23), 4337–4340.
182. Li, W. J., Chou, S. L., Wang, J. Z., Liu, H. K., Dou, S. X. A new, cheap, and productive FeP anode material for sodium-ion batteries (vol 51, pg 3682, 2015). *Chem. Commun.* **2015**, *51* (22), 4720–4720.
183. Kim, Y., Kim, Y., Choi, A., Woo, S., Mok, D., Choi, N. S., Jung, Y. S., Ryu, J. H., Oh, S. M., Lee, K. T. Tin phosphide as a promising anode material for Na-ion batteries. *Adv. Mater.* **2014**, *26* (24), 4139–4144.
184. Phil, L., Naveed, M., Mohammad, I. S., Bo, L., Bin, D. Chitooligosaccharide: An evaluation of physicochemical and biological properties with the proposition for determination of thermal degradation products. *Biomed. Pharmacother.* **2018**, *102*, 438–451.
185. Li, W. J., Yang, Q. R., Chou, S. L., Wang, J. Z., Liu, H. K. Cobalt phosphide as a new anode material for sodium storage. *J. Power Sources* **2015**, *294*, 627–632.
186. Zhang, W. J., Dahbi, M., Amagasa, S., Yamada, Y., Komaba, S. Iron phosphide as negative electrode material for Na-ion batteries. *Electrochem. Commun.* **2016**, *69*, 11–14.
187. Lee, J. H., Yoon, C. S., Hwang, J. Y., Kim, S. J., Maglia, F., Lamp, P., Myung, S. T., Sun, Y. K. High-energy-density lithium-ion battery using a carbon-nanotube-Si composite anode and a compositionally graded Li[Ni0.85Co0.05Mn0.10]O-2 cathode. *Energ. Environ. Sci.* **2016**, *9* (6), 2152–2158.
188. Wang, Y. X., Yang, J. P., Chou, S. L., Liu, H. K., Zhang, W. X., Zhao, D. Y., Dou, S. X. Uniform yolk-shell iron sulfide-carbon nanospheres for superior sodium-iron sulfide batteries. *Nat. Commun.* **2015**, *6*.

189. Ryu, H. S., Kim, J. S., Park, J. S., Park, J. W., Kim, K. W., Ahn, J. H., Nam, T. H., Wang, G. X., Ahn, H. J. Electrochemical properties and discharge mechanism of Na/TiS$_2$ cells with liquid electrolyte at room temperature. *J. Electrochem. Soc.* **2013**, *160* (2), A338–A343.
190. Wang, Q. H., Jiao, L. F., Du, H. M., Peng, W. X., Han, Y., Song, D. W., Si, Y. C., Wang, Y. J., Yuan, H. T. Novel flower-like CoS hierarchitectures: One-pot synthesis and electrochemical properties. *J. Mater. Chem.* **2011**, *21* (2), 327–329.
191. Peng, S. J., Han, X. P., Li, L. L., Zhu, Z. Q., Cheng, F. Y., Srinivansan, M., Adams, S., Ramakrishna, S. Unique cobalt sulfide/reduced graphene oxide composite as an anode for sodium-ion batteries with superior rate capability and long cycling stability. *Small* **2016**, *12* (10), 1359–1368.
192. Zhang, L. S., Zuo, L. Z., Fan, W., Liu, T. X. NiCo$_2$S$_4$ nanosheets grown on 3D networks of nitrogen-doped graphene/carbon nanotubes: Advanced anode materials for lithium-ion batteries. *Chemelectrochem* **2016**, *3* (9), 1384–1391.
193. Fan, X. L., Mao, J. F., Zhu, Y. J., Luo, C., Suo, L. M., Gao, T., Han, F. D., Liou, S. C., Wang, C. S. Superior stable self-healing SnP$_3$ anode for sodium-ion batteries. *Adv. Energy Mater.* **2015**, *5* (18).
194. Mortazavi, M., Ye, Q. J., Birbilis, N., Medhekar, N. V. High capacity group-15 alloy anodes for Na-ion batteries: Electrochemical and mechanical insights. *J.Power Sources* **2015**, *285*, 29–36.
195. Zhang, C., Wang, X., Liang, Q. F., Liu, X. Z., Weng, Q. H., Liu, J. W., Yang, Y. J. et al. Amorphous phosphorus/nitrogen-doped graphene paper for ultrastable sodium-ion batteries. *Nano Lett.* **2016**, *16* (3), 2054–2060.
196. Morito, H., Yamada, T., Ikeda, T., Yamane, H. Na-Si binary phase diagram and solution growth of silicon crystals. *J. Alloy Compd.* **2009**, *480* (2), 723–726.
197. Xu, Y. L., Swaans, E., Basak, S., Zandbergen, H. W., Borsa, D. M., Mulder, F. M. Reversible Na-ion uptake in Si nanoparticles. *Adv. Energy Mater.* **2016**, *6* (2).
198. Zhang, L., Hu, X. L., Chen, C. J., Guo, H. P., Liu, X. X., Xu, G. Z., Zhong, H. J. et al. In operando mechanism analysis on nanocrystalline silicon anode material for reversible and ultrafast sodium storage. *Adv. Mater.* **2017**, *29* (5).
199. Zhang, L., Rajagopalan, R., Guo, H. P., Hu, X. L., Dou, S. X., Liu, H. K. A green and facile way to prepare granadilla-like silicon-based anode materials for Li-ion batteries. *Adv. Funct. Mater.* **2016**, *26* (3), 440–446.
200. Sangster, J., Pelton, A. D. The Ge-Na (germanium-sodium) system. *J. Phase Equilib.* **1997**, *18* (3), 295–297.
201. Yue, C., Yu, Y. J., Sun, S. B., He, X., Chen, B. B., Lin, W., Xu, B. B. et al. High performance 3D Si/Ge nanorods array anode buffered by TiN/Ti interlayer for sodium-ion batteries. *Adv. Funct. Mater.* **2015**, *25* (9), 1386–1392.
202. Abel, P. R., Lin, Y. M., de Souza, T., Chou, C. Y., Gupta, A., Goodenough, J. B., Hwang, G. S., Heller, A., Mullins, C. B., Nanocolumnar germanium thin films as a high-rate sodium-ion battery anode material. *J. Phys. Chem. C* **2013**, *117* (37), 18885–18890.
203. Baggetto, L., Keum, J. K., Browning, J. F., Veith, G. M. Germanium as negative electrode material for sodium-ion batteries. *Electrochem. Commun.* **2013**, *34*, 41–44.
204. Komaba, S., Matsuura, Y., Ishikawa, T., Yabuuchi, N., Murata, W., Kuze, S. Redox reaction of Sn-polyacrylate electrodes in aprotic Na cell. *Electrochem. Commun.* **2012**, *21*, 65–68.
205. Wang, J. W., Liu, X. H., Mao, S. X., Huang, J. Y. Microstructural evolution of tin nanoparticles during *in situ* sodium insertion and extraction. *Nano Lett.* **2012**, *12* (11), 5897–5902.

206. Wang, J. J., Eng, C., Chen-Wiegart, Y. C. K., Wang, J. Probing three-dimensional sodiation-desodiation equilibrium in sodium-ion batteries by in situ hard X-ray nanotomography. *Nat. Commun.* **2015**, 6.
207. Bresser, D., Mueller, F., Buchholz, D., Paillard, E., Passerini, S. Embedding tin nanoparticles in micron-sized disordered carbon for lithium- and sodium-ion anodes. *Electrochim. Acta* **2014**, *128*, 163–171.
208. Liu, Y. C., Zhang, N., Jiao, L. F., Tao, Z. L., Chen, J. Ultrasmall Sn nanoparticles embedded in carbon as high-performance anode for sodium-ion batteries. *Adv. Funct. Mater.* **2015**, *25* (2), 214–220.
209. Liu, Y. C., Zhang, N., Jiao, L. F., Chen, J. Tin nanodots encapsulated in porous nitrogen-doped carbon nanofibers as a free-standing anode for advanced sodium-ion batteries. *Adv. Mater.* **2015**, *27* (42), 6702.
210. Luo, B., Qiu, T. F., Ye, D. L., Wang, L. Z., Zhi, L. J. Tin nanoparticles encapsulated in graphene backboned carbonaceous foams as high-performance anodes for lithium-ion and sodium-ion storage. *Nano Energy* **2016**, *22*, 232–240.
211. Ellis, L. D., Hatchard, T. D., Obrovac, M. N. Reversible insertion of sodium in tin. *J. Electrochem. Soc.* **2012**, *159* (11), A1801–A1805.
212. Sangster, J. Thermodynamics and phase diagrams of 32 binary common-ion systems of the group Li, Na, K,Rb, Cs//F, Cl, Br, I,OH, NO$_3$. *J. Phase Equilib.* **2000**, *21* (3), 241–268.
213. Darwiche, A., Marino, C., Sougrati, M. T., Fraisse, B., Stievano, L., Monconduit, L. Better cycling performances of bulk Sb in Na-ion batteries compared to Li-ion systems: An unexpected electrochemical mechanism (vol 134, pg 20805, 2012). *J. Am. Chem. Soc.* **2013**, *135* (27), 10179–10179.
214. Qian, J. F., Chen, Y., Wu, L., Cao, Y. L., Ai, X. P., Yang, H. X. High capacity Na-storage and superior cyclability of nanocomposite Sb/C anode for Na-ion batteries. *Chem. Commun.* **2012**, *48* (56), 7070–7072.
215. Darwiche, A., Marino, C., Sougrati, M. T., Fraisse, B., Stievano, L., Monconduit, L. Better cycling performances of bulk Sb in Na-ion batteries compared to Li-ion systems: An unexpected electrochemical mechanism. *J. Am. Chem. Soc.* **2012**, *134* (51), 20805–20811.
216. Xiao, L. F., Cao, Y. L., Xiao, J., Wang, W., Kovarik, L., Nie, Z. M., Liu, J. High capacity, reversible alloying reactions in SnSb/C nanocomposites for Na-ion battery applications. *Chem. Commun.* **2012**, *48* (27), 3321–3323.
217. Zhou, X. S., Dai, Z. H., Bao, J. C., Guo, Y. G. Wet milled synthesis of an Sb/MWCNT nanocomposite for improved sodium storage. *J. Mater. Chem. A* **2013**, *1* (44), 13727–13731.
218. He, M., Kraychyk, K., Walter, M., Kovalenko, M. V. Monodisperse antimony nanocrystals for high-rate Li-ion and Na-ion battery anodes: Nano versus bulk. *Nano Lett.* **2014**, *14* (3), 1255–1262.
219. Hou, H. S., Yang, Y. C., Zhu, Y. R., Jing, M. J., Pan, C. C., Fang, L. B., Song, W. X., Yang, X. M., Ji, X. B. An electrochemical study of Sb/acetylene black composite as anode for sodium-ion batteries. *Electrochim. Acta* **2014**, *146*, 328–334.
220. Zhou, X. L., Zhong, Y. R., Yang, M., Hu, M., Wei, J. P., Zhou, Z. Sb nanoparticles decorated N-rich carbon nanosheets as anode materials for sodium ion batteries with superior rate capability and long cycling stability. *Chem. Commun.* **2014**, *50* (85), 12888–12891.

221. Zhang, N., Liu, Y. C., Lu, Y. Y., Han, X. P., Cheng, F. Y., Chen, J. Spherical nano-Sb@C composite as a high-rate and ultra-stable anode material for sodium-ion batteries. *Nano Res.* **2015**, *8* (10), 3384–3393.
222. Zhu, Y. J., Han, X. G., Xu, Y. H., Liu, Y. H., Zheng, S. Y., Xu, K., Hu, L. B., Wang, C. S. Electrospun Sb/C fibers for a stable and fast sodium-ion battery anode. *ACS Nano* **2013**, *7* (7), 6378–6386.
223. Hou, H. S., Jing, M. J., Yang, Y. C., Zhang, Y., Song, W. X., Yang, X. M., Chen, J., Chen, Q. Y., Ji, X. B. Antimony nanoparticles anchored on interconnected carbon nanofibers networks as advanced anode material for sodium-ion batteries. *J. Power Sources* **2015**, *284*, 227–235.
224. Wu, L., Hu, X. H., Qian, J. F., Pei, F., Wu, F. Y., Mao, R. J., Ai, X. P., Yang, H. X., Cao, Y. L. Sb-C nanofibers with long cycle life as an anode material for high-performance sodium-ion batteries. *Energ. Environ. Sci.* **2014**, *7* (1), 323–328.
225. Nithya, C., Gopukumar, S. rGO/nano Sb composite: A high performance anode material for Na+ ion batteries and evidence for the formation of nanoribbons from the nano rGO sheet during galvanostatic cycling. *J. Mater. Chem. A* **2014**, *2* (27), 10516–10525.
226. Hu, L. Y., Zhu, X. S., Du, Y. C., Li, Y. F., Zhou, X. S., Bao, J. C. A chemically coupled antimony/multilayer graphene hybrid as a high-performance anode for sodium-ion batteries. *Chem. Mater.* **2015**, *27* (23), 8138–8145.
227. Kim, Y., Park, Y., Choi, A., Choi, N. S., Kim, J., Lee, J., Ryu, J. H., Oh, S. M., Lee, K. T. An amorphous red phosphorus/carbon composite as a promising anode material for sodium ion batteries. *Adv. Mater.* **2013**, *25* (22), 3045–3049.
228. Sangster, J. M. Na-P (sodium-phosphorus) system. *J. Phase Equilib. Diff.* **2010**, *31* (1), 62–67.
229. Sun, J., Lee, H. W., Pasta, M., Yuan, H. T., Zheng, G. Y., Sun, Y. M., Li, Y. Z., Cui, Y. A phosphorene-graphene hybrid material as a high-capacity anode for sodium-ion batteries. *Nat. Nanotechnol.* **2015**, *10* (11), 980.
230. Song, J. X., Yu, Z. X., Gordin, M. L., Hu, S., Yi, R., Tang, D. H., Walter, T. et al. Chemically bonded phosphorus/graphene hybrid as a high performance anode for sodium-ion batteries. *Nano Lett.* **2014**, *14* (11), 6329–6335.
231. Qian, J. F., Wu, X. Y., Cao, Y. L., Ai, X. P., Yang, H. X. High capacity and rate capability of amorphous phosphorus for sodium ion batteries. *Angew. Chem. Int. Edit.* **2013**, *52* (17), 4633–4636.
232. Sun, J., Zheng, G. Y., Lee, H. W., Liu, N., Wang, H. T., Yao, H. B., Yang, W. S., Cui, Y. Formation of stable phosphorus-carbon bond for enhanced performance in black phosphorus nanoparticle-graphite composite battery anodes. *Nano Lett.* **2014**, *14* (8), 4573–4580.
233. Ramireddy, T., Xing, T., Rahman, M. M., Chen, Y., Dutercq, Q., Gunzelmann, D., Glushenkov, A. M. Phosphorus-carbon nanocomposite anodes for lithium-ion and sodium-ion batteries. *J. Mater. Chem. A* **2015**, *3* (10), 5572–5584.
234. Dahbi, M., Yabuuchi, N., Fukunishi, M., Kubota, K., Chihara, K., Tokiwa, K., Yu, X. F. et al. Black phosphorus as a high-capacity, high-capability negative electrode for sodium-ion batteries: Investigation of the electrode/interface. *Chem. Mater.* **2016**, *28* (6), 1625–1635.
235. Liu, J., Wen, Y. R., van Aken, P. A., Maier, J., Yu, Y. Facile synthesis of highly porous Ni-Sn intermetallic microcages with excellent electrochemical performance for lithium and sodium storage. *Nano Lett.* **2014**, *14* (11), 6387–6392.

236. Baggetto, L., Hah, H. Y., Johnson, C. E., Bridges, C. A., Johnson, J. A., Veith, G. M. The reaction mechanism of FeSb$_2$ as anode for sodium-ion batteries. *Phys. Chem. Chem. Phys.* **2014**, *16* (20), 9538–9545.
237. Jackson, E. D., Green, S., Prieto, A. L. Electrochemical performance of electrodeposited Zn$_4$Sb$_3$ films for sodium-ion secondary battery anodes. *ACS Appl. Mater. Inter.* **2015**, *7* (14), 7447–7450.
238. Baggetto, L., Carroll, K. J., Hah, H. Y., Johnson, C. E., Mullins, D. R., Unocic, R. R., Johnson, J. A., Meng, Y. S., Veith, G. M. Probing the mechanism of sodium ion insertion into copper antimony Cu$_2$Sb anodes. *J. Phys. Chem. C* **2014**, *118* (15), 7856–7864.
239. Lin, Y. M., Abel, P. R., Gupta, A., Goodenough, J. B., Heller, A., Mullins, C. B. Sn-Cu nanocomposite anodes for rechargeable sodium-ion batteries. *ACS Appl. Mater. Inter.* **2013**, *5* (17), 8273–8277.
240. Baggetto, L., Allcorn, E., Unocic, R. R., Manthiram, A., Veith, G. M. Mo3Sb7 as a very fast anode material for lithium-ion and sodium-ion batteries. *J. Mater. Chem. A* **2013**, *1* (37), 11163–11169.
241. Zhao, Y. B., Manthiram, A. High-capacity, high-rate Bi-Sb alloy anodes for lithium-ion and sodium-ion batteries. *Chem. Mater.* **2015**, *27* (8), 3096–3101.
242. Kim, I. T., Kim, S. O., Manthiram, A. Effect of TiC addition on SnSb-C composite anodes for sodium-ion batteries. *J. Power Sources* **2014**, *269*, 848–854.
243. Ji, L. W., Zhou, W. D., Chabot, V., Yu, A. P., Xiao, X. C. Reduced graphene oxide/tin-antimony nanocomposites as anode materials for advanced sodium-ion batteries. *ACS Appl. Mater. Inter.* **2015**, *7* (44), 24895–24901.
244. Hassoun, J., Panero, S., Simon, P., Taberna, P. L., Scrosati, B. High-rate, long-life Ni-Sn nanostructured electrodes for lithium-ion batteries. *Adv. Mater.* **2007**, *19* (12), 1632.
245. Haupler, B., Wild, A., Schubert, U. S. Carbonyls: Powerful organic materials for secondary batteries. *Adv. Energy Mater.* **2015**, *5* (11).
246. Zhao, L., Zhao, J. M., Hu, Y. S., Li, H., Zhou, Z. B., Armand, M., Chen, L. Q. Disodium terephthalate (Na$_2$C$_8$H$_4$O$_4$) as high performance anode material for low-cost room-temperature sodium-ion battery. *Adv. Energy Mater.* **2012**, *2* (8), 962–965.
247. Park, Y., Shin, D. S., Woo, S. H., Choi, N. S., Shin, K. H., Oh, S. M., Lee, K. T., Hong, S. Y. Sodium terephthalate as an organic anode material for sodium ion batteries. *Adv. Mater.* **2012**, *24* (26), 3562–3567.
248. Abouimrane, A., Weng, W., Eltayeb, H., Cui, Y. J., Niklas, J., Poluektov, O., Amine, K. Sodium insertion in carboxylate based materials and their application in 3.6 V full sodium cells. *Energ. Environ. Sci.* **2012**, *5* (11), 9632–9638.
249. Deng, W. W., Liang, X. M., Wu, X. Y., Qian, J. F., Cao, Y. L., Ai, X. P., Feng, J. W., Yang, H. X. A low cost, all-organic Na-ion battery based on polymeric cathode and anode. *Sci. Rep.-UK* **2013**, *3*.
250. Wang, Y., Kretschmer, K., Zhang, J. Q., Mondal, A. K., Guo, X., Wang, G. X. Organic sodium terephthalate@graphene hybrid anode materials for sodium-ion batteries. *RSC Adv.* **2016**, *6* (62), S7098–S7102.
251. Choi, A., Kim, Y. K., Kim, T. K., Kwon, M. S., Lee, K. T., Moon, H. R. 4,4'-Biphenyldicarboxylate sodium coordination compounds as anodes for Na-ion batteries. *J. Mater. Chem. A* **2014**, *2* (36), 14986–14993.
252. Wang, H. G., Yuan, S., Si, Z. J., Zhang, X. B. Multi-ring aromatic carbonyl compounds enabling high capacity and stable performance of sodium-organic batteries. *Energ. Environ. Sci.* **2015**, *8* (11), 3160–3165.

253. Deng, W. W., Qian, J. F., Cao, Y. L., Ai, X. P., Yang, H. X. Graphene-wrapped Na$_2$C$_{12}$H$_6$O$_4$ nanoflowers as high performance anodes for sodium-ion batteries. *Small* **2016**, *12* (5), 583–587.
254. Wang, C. L., Xu, Y., Fang, Y. G., Zhou, M., Liang, L. Y., Singh, S., Zhao, H. P., Schober, A., Lei, Y. Extended pi-conjugated system for fast-charge and -discharge sodium-ion batteries. *J. Am. Chem. Soc.* **2015**, *137* (8), 3124–3130.
255. Zhao, R. R., Cao, Y. L., Ai, X. P., Yang, H. X. Reversible Li and Na storage behaviors of perylenetetracarboxylates as organic anodes for Li- and Na-ion batteries. *J. Electroanal Chem.* **2013**, *688*, 93–97.
256. Wang, S. W., Wang, L. J., Zhu, Z. Q., Hu, Z., Zhao, Q., Chen, J. All organic sodium-ion batteries with Na$_4$C$_8$H$_2$O$_6$. *Angew. Chem. Int. Edit.* **2014**, *53* (23), 5892–5896.
257. Renault, S., Mihali, V. A., Edstrom, K., Brandell, D. Stability of organic Na-ion battery electrode materials: The case of disodium pyromellitic diimidate. *Electrochem. Commun.* **2014**, *45*, 52–55.
258. Li, Z. T., Zhou, J. Y., Xu, R. F., Liu, S. P., Wang, Y. K., Li, P., Wu, W. T., Wu, M. B. Synthesis of three dimensional extended conjugated polyimide and application as sodium-ion battery anode. *Chem. Eng. J.* **2016**, *287*, 516–522.
259. Chen, L., Li, W. Y., Wang, Y. G., Wang, C. X., Xia, Y. Y. Polyimide as anode electrode material for rechargeable sodium batteries. *RSC Adv.* **2014**, *4* (48), 25369–25373.
260. Luo, C., Zhu, Y. J., Xu, Y. H., Liu, Y. H., Gao, T., Wang, J., Wang, C. S. Graphene oxide wrapped croconic acid disodium salt for sodium ion battery electrodes. *J. Power Sources* **2014**, *250*, 372–378.
261. Wang, H., Hu, P. F., Yang, J., Gong, G. M., Guo, L., Chen, X. D. Renewable-juglone-based high-performance sodium-ion batteries. *Adv. Mater.* **2015**, *27* (14), 2348–2354.
262. Zhu, Z. Q., Li, H., Liang, J., Tao, Z. L., Chen, J. The disodium salt of 2,5-dihydroxy-1,4-benzoquinone as anode material for rechargeable sodium ion batteries. *Chem. Commun.* **2015**, *51* (8), 1446–1448.
263. Zhu, H., Yin, J., Zhao, X., Wang, C. Y., Yang, X. R. Humic acid as promising organic anodes for lithium/sodium ion batteries. *Chem. Commun.* **2015**, *51* (79), 14708–14711.
264. Luo, C., Wang, J. J., Fan, X. L., Zhu, Y. J., Han, F. D., Suo, L. M., Wang, C. S. Roll-to-roll fabrication of organic nanorod electrodes for sodium ion batteries. *Nano Energy* **2015**, *13*, 537–545.

4

Cathode Materials for Sodium Ion Batteries

4.1 Introduction

Both sodium ion batteries (SIBs) and lithium ion batteries (LIBs) were invented around the same period of time, however, the remarkable electrochemical achievement and rapid commercialization of LIBs have interrupted the research on SIBs. But in recent years the trend has been changing where many academic and industrial researches have shown proven potential for SIBs from a commercialization aspect. Companies like Faradion, United Kingdom, and Aquion Energy, United States, are extensively working on full-cell SIBs for practical applications. For instance, as a proof of concept, Faradion demonstrated fully functional e-bikes based on SIBs. This has opened much room to explore and optimize a new set of electrode systems for SIBs for practical applications. In the current scenario, the SIB technology is driven by the performance of the cathode materials. Thus, the first challenge is to develop potential cathode materials. However, innovative development of appropriate cathode materials which exhibit multiple characteristics, such as high specific capacity, superior cycling stability, and good rate capability could meet many obstacles. For instance, one of the major obstacles is that the larger Na^+ radius (0.98 Å) than that of Li^+ (0.69 Å) leads to the sluggish kinetics during the Na^+ insertion/extraction process and transport across the host-material framework, this will always cause degradation in specific capacity and rate capacity.[1,2] Another drawback is that, due to the bulky nature of the Na^+, during the insertion mechanism larger volume expansion could occur, this could cause a change in the phase and lattice of the host materials, resulting in deterioration in the cycling performance. Another major problem is the lower specific energy compared to the LIBs, because of the lower operating potential and larger atomic weight of sodium. Thus, many recent studies have tried to overcome all these disadvantages by exploring and exploiting the potential properties of sodium-based cathodes. Further, many researches focused on unraveling the detailed interaction mechanism of sodium ion with the cathode materials to have an in-depth understanding of the atomic level sodium transports and interactions.[3,4] This rigorous understanding could offer answers to the existing problems by giving valuable insight on the structural modification of the materials to provide better electrochemical performance.[5] Thus, in this chapter, recent progress in the cathode materials are organized to get a wider and better perspective in the area of SIBs. Furthermore, the fundamental aspects of SIBs, processing, and characterization techniques are also discussed. Some inherent properties of many well-studied cathodes (layered transition metal oxides, phosphates, pyrophosphate, Sodium super ion conductor (NASICON) -based compounds, fluorides, and organic compounds) for SIBs are also analyzed in this chapter. Through the in-depth understanding of all the abovementioned topics, the causes behind the drawbacks of the existing cathode systems, and how these issues can be addressed by designing new cathodes with improved electrochemical performance, are revealed and explained. In the summary, challenges and future prospects are also incorporated to provide insight and guide the readers to design and develop new cathode systems with enhanced electrochemical properties for SIBs.

4.2 Transition Metal Oxides

4.2.1 Layered Sodium Metal Oxides

4.2.1.1 Sodium-Based Single Transition-Metal Oxides

Layered cathode materials for sodium ion storage were extensively studied during the 1970s and 1980s. The common chemical formula for these sodium metal oxides is Na_xMO_{2+y}, where M is a transition metal.[6-9] In general, the packing of the cathode systems takes place in two ways: the O3 and P2 types, where O and P demonstrate the octahedral and trigonal prismatic coordinations for sodium ions and 3 or 2 corresponds to the number of transition metal layers in a repeated stacking unit (Figure 4.1a).[10] Initially, like LIBs, the studies on SIBs were also focused on sodium cobalt- and sodium manganese oxides-based materials. In general, these crystals are formed from sheets of edge-sharing octahedral MO_6. Here, the Na ions are placed between these octahedral sheets. The Na analogue of these kinds of crystal structures exhibits an electrochemical insertion mechanism similar to that of Li cathodes. However, during the de-insertion process of Na ions from the alkali ion sheets, these types of materials showed many complex phase transitions.[11] This could be attributed to multiple factors, such as the larger Na ion radius, the ordering arrangement between Na^+ and V_{Na}^+, the elongated Na-O bonds, etc.[12] Another major concern of these types of layered oxides is their poor air stability. For instance, O3-type materials can provide higher reversible capacities; however, they suffer from poor air stability and low cycling performance. On the other hand, P2-type cathodes can have superior cycling performance and air stability. This is mainly because of the occupation of Na^+ ions in the larger trigonal prismatic sites, which could help the transportation of Na+ ions.[13–16] In recent years, many layered sodium transition oxides with a common chemical formula of Na$_x$MO$_2$ (M = Ni, Co, Mn, Fe, Cr, V, etc.) have been explored as cathode material and have been observed to show reversible insertion/de-insertion of Na+ ions. Owing to the success of LiCoO$_2$ in LIBs, the NaCoO$_2$ has been widely investigated as a cathode material for SIBs. But the larger size of the Na ion restricted the practical capacity to less than 90 mA h g^{-1}. Thus, many off-stoichiometric Na$_x$CoO$_2$, such as O3- (0.83 < x < 1.0) and P2- (0.67 < x < 0.80) type compounds, have been explored to meet the poor electrochemical properties of the NaCoO$_2$ materials. Even though O3 phase showed better ion diffusion coefficients as compared to P2 phase, but as mentioned above, the O3-type material suffered from poor cycling performance because of the phase transitions (O3–O'3–P3–P2), which leads to higher polarization, because of the layer gliding at room temperature with the variation in Na concentration. On the other hand, the P2 phase can be protected when starting with P2-type Na$_x$CoO$_2$ material, during the insertion/extraction process, resulting in a low polarization and hence with improved structural stability and superior cycling performance.[17] The sodium diffusion pathway in a different-layered structure is shown in Figure 4.1b. Since the local environment of Na ions at prismatic sites is different, the diffusion pathway will also be different as shown in Figure 4.1b.[10] Another potential cathode material is NaMnO$_2$. This material is considered to be more promising than NaCoO$_2$ because of its superior properties, such as high theoretical capacity (242 mA h g^{-1}), low-cost, material abundance, and high activity of M^{4+}/M^{3+} redox couple. Apart from the common O3 and P2 phases; the O2 phase and birnessite phase also demonstrate potential electrochemical properties. The O2 crystal structure is

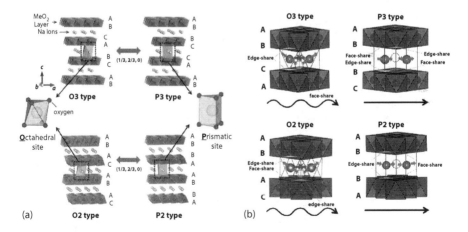

FIGURE 4.1
(a) Different classification of layered materials and (b) Na migration paths in different layered structures. (Reprinted with permission from Yabuuchi, N. et al., *Chem. Rev.*, 114, 11636, 2014. Copyright 2014 American Chemical Society).

generated from zigzag layers of edge-shared MnO_6 octahedral structure and Na^+ ions occupied in the octahedral sites. The birnessite structure is formed by lamellar arrangement of synthetic nanosheets, which can be synthesized by a hydrothermal reaction from the precursors, such as NaOH, H_2O_2, and $Mn(NO_3)_2$. Unlike the lithium counterparts, all the layered $NaMnO_2$ cathode systems demonstrate substantial intercalation mechanisms and can maintain sodium de-intercalation without any phase conversion to the spinel structure. However, in contrast to layered $LiMnO_2$, Na_xMnO_2 experiences low cycling stability due to its structural disintegration and amorphization caused by constant strains and distortions in the crystal structure. For instance, birnessite-structured ($NaMn_3O_5$) materials can deliver a large discharge capacity of 219 mA h g^{-1} in the potential window of 1.5–4.7 V, but it could only retain 70% of its initial capacity at the end of 20 cycles.[18] But β-$NaMnO_2$ crystal phases demonstrated stable, reproducible, and reversible Na intercalation chemistry, with >70% initial capacity retention after 100 cycles of charging/discharging.[19] Due to the low-cost and availability, the iron-based cathodes, such as O3-$NaFeO_2$ have attracted considerable attention as potential cathode materials for SIBs.[20-23] Many studies have revealed, however, that the Li analogue O3-$LiFeO_2$ is electrochemically inactive due to the difficulty in getting stable Fe^{4+} by the oxidization of Fe^{3+}.[23] For these materials, the obtained charge capacity is mainly attributed to the dominant oxygen removal process at the solid electrolyte interphase rather than Fe^{4+}/Fe^{3+} redox reaction. This is because the Fe^{3+} 3d orbital is strongly hybridized with the 2p oxygen orbital in the system, so that oxygen removal is more favorable than the oxidation of Fe^{3+} to Fe^{4+}. On the other hand, the α-$NaFeO_2$ showed good reversible Na ion intercalation chemistry. Apart from this, α-$NaFeO_2$ material is more economic, earth-abundant, and environmentally friendly in nature due to the presence of iron. The high theoretical capacity of 242 mA h g^{-1} of Fe^{3+}/Fe^{4+} redox couple also makes this system a promising candidate. Nonetheless, most of these $NaFeO_2$ materials show low experimental capacity of ~85 mA h g^{-1} with a voltage plateau at 3.3 V, perhaps due to the Fe^{4+} instability. A higher operating voltage of >3.5 V causes poor rate capability and loss of a sodium transition pathway, possibly due to the irreversible structural degradation resulting from the Jahn-Teller distortion and polarization.[24] However, due to the superior thermal stability of α-$NaFeO_2$ as compared to its lithium counterpart, which could apply in large scale applications like electric vehicles (EVs) to avoid serious security issues after meeting the required electrochemical performance by further intense research and optimization.[21] Thus, many studies have tried to enhance the electrochemical properties of $NaFeO_2$ cathode materials. For instance, a study showed that polypyrrole coating can direct an improved capacity of 120 mA h g^{-1} at C/10 even at the end of 100 cycles in $NaFeO_2$.[25] Another notable cathode for SIBs is $NaCrO_2$, but the Li analogue of $NaCrO_2$ ($R\bar{3}m$) is electrochemically inactive.[26-29] These $NaCrO_2$ cathode structures could exhibit a high theoretical capacity of 250 mA h g^{-1} based on the Cr^{6+}/Cr^{3+} redox couple. $Na_{0.5}CrO_2$

could show enhanced safety characteristics and could demonstrate superior thermal stability to commercially available $Na_{0.5}CoO_2$.[30] However, these $NaCrO_2$ cathodes experience large capacity fading when the Na extraction exceeds particular capacity limits. That is, the Na_xCrO_2 (0 < x < 1) could maintain its layered crystal structure when x < 0.4 with an O3–P3 phase transition. But a further Na extraction could cause a layered-to-rock-salt (CrO_2) phase transformation, which could introduce a large polarization (Figure 4.2a). Thus, it is critical to avoid this layered-to-rock-salt phase change to get enhanced electrochemical performance, mainly the stability. One proven technique to enhance the electrochemical performance is to wrap this O3-type $NaCrO_2$ using a thin open framework of carbon. For instance, a 10 nm carbon-layer wrapped $NaCrO_2$ cathode system showed an excellent rate performance of up to 150 charge/discharge cycles and good cycling performance after 300 cycles of consecutive charge/discharge.[31] Another potential method is to reduce the concentration of Cr^{4+} triplets (in order to minimize Cr^4 dis-proportionation) via chemical substitutions of other cations to form multi-cationic transition metal oxides.[32] The $NaNiO_2$ material with space group of $C2/m$ has been demonstrated as a potential cathode for SIBs.

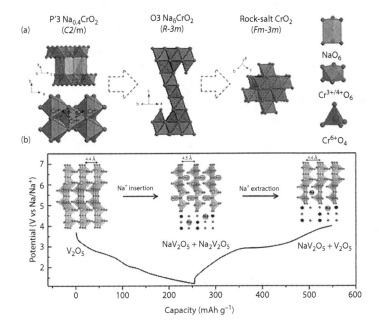

FIGURE 4.2
(a) Crystal structure of Na_xCrO_2. (Reprinted with permission from Bo, S.-H. et al., *Chem. Mater.*, 28, 1419, 2016 <https://pubs.acs.org/doi/pdf/10.1021/acs.chemmater.5b04626>. Copyright 2016 American Chemical Society.) (Further permission to use this figure should be directed to the ACS) and (b) Crystal structure and intercalation/de-intercalation channels along with charge-discharge process of V_2O_5. (Reprinted with permission from Ali, G. et al., *ACS Appl. Mater. Interfaces*, 8, 6032, 2016. Copyright 2016 American Chemical Society.)

But the initial study on this material showed that only 0.2 Na could be extracted from the monoclinic crystal structure.[33,34] Later studies re-estimated the Na storage capacity of NaNiO$_2$, however, and they have demonstrated a capacity of 123 mA h g^{-1}, which is equivalent to 0.52 Na.[35] Apart from all these aforementioned-layered Na-based transition oxide cathodes, other materials such as NaCrO$_2$,[27,32] NaNbO$_2$,[36] Na$_2$RuO$_3$,[37] and NaV$_x$O$_y$,[7,38,39] have also been extensively investigated to achieve superior electrochemical properties. The abovementioned single-transition sodium-based oxides can be easily processed under normal atmosphere via different facile methods such as solid-state technique, co-precipitation reaction, and hydrothermal process. Synthesis of these materials in low volume or from low concentrated precursors could show improved electrochemical performances such as capacity, rate capability, etc., as compared to bulk scale materials.

4.2.1.2 Sodium-Based Mixed-Cation Oxides

Exploiting the synergistic contributions of different metals could help to enhance the electrochemical performance as well as the structural stability of the cathode materials. Other than the enhanced electrochemical properties and stability; these multi-cationic compounds could be beneficial in terms of cost reduction and environmental benignity, by partially replacing the expensive metal ions like Ni or Co with less expensive and environmentally friendly materials without compromising electrochemical performance and structural stability. Material with various active redox couples like Ni^{2+}/Ni^{3+}/Ni^{4+}, Co^{3+}/Co^{4+}, and Fe^{3+}/Fe^{4+} can participate in the redox process and can contribute to enhance the specific capacity and other electrochemical properties. Several redox-inactive cations such as Mn^{4+} function as a structural noose to stabilize the lattice structure of the layered compounds. For example, materials like P2-Na$_{2/3}$[Ni$_{1/3}$Mn$_{2/3}$]O$_2$ have been considered as a potential multi-cationic-layered oxide for SIBs. This material has Ni^{2+} redox active ions and Mn^{4+} redox inactive ions, which exhibited excellent reversibility in the potential window of 2.0–4.0 V with a specific capacity of 86 mA h g^{-1} at 0.1C rate.[40] Apart from that, this material maintained its P2 phase even at the end of long consecutive charge/discharge cycles in the same potential range.[41] Another similar compound is P2-Na$_{0.67}$Co$_{0.5}$Mn$_{0.5}$O$_2$, which could deliver an excellent specific capacity [147 mA h g^{-1} (0.1C)] and outstanding cyclic performance (~100% capacity retention at the end of 100 charge/discharge cycles (1C)] with good structural stability.[42] In this compound, the doped Mn helped to minimize the Jahn-Teller-induced crystal structural transition and ordering processes of Na$^+$; which lead to better cycling stability through improved Na$^+$ kinetics. Another high capacity multi-cationic compound is P2-Na$_{2/3}$(Fe$_{1/2}$Mn$_{1/2}$)O$_2$. In a recent study, the hierarchical nanofibers of P2-Na$_{2/3}$(Fe$_{1/2}$Mn$_{1/2}$)O$_2$ were fabricated through electrospinning techniques. The electrospun nanofibers exhibited a specific capacity of ~195 mA h g^{-1} (0.1C) with >86% initial capacity retention at the end of 80 charge/discharge cycles.[43] O3-Na$_{0.9}$[Cu$_{0.22}$Fe$_{0.30}$Mn$_{0.48}$]O$_2$, Co/Ni-free cathode materials showed stability in air. This material demonstrated a specific energy density of 210 W h kg^{-1} in the potential range of 2.5–4.5 V.[16] Further, to improve the electrochemical properties of the SIBs, many studies have conducted and come up with new ternary and quaternary multi-cationic cathodes.[44–46] For instance, P2-Na$_{0.5}$[Ni$_{0.23}$Fe$_{0.13}$Mn$_{0.63}$]O$_2$ showed good cycling performance at higher C-rates (95.3% capacity retention at the end of 40 cycles at 5C rate) and excellent specific discharge capacity (200 mA h g^{-1} in the wide potential window of 4.6–1.5 V).[47] Yet another promising cathode material is Na[Ni$_{0.60}$Co$_{0.05}$Mn$_{0.35}$]O$_2$. The nanostructure hierarchical columnar Na[Ni$_{0.60}$Co$_{0.05}$Mn$_{0.35}$]O$_2$ showed a specific discharge capacity of 157 mA h g^{-1} at a current density of 15 mA g^{-1} and retained 80% of its initial capacity over 100 cycles. This particular compound also demonstrated promising full-cell performance when tested against a hard carbon anode. In the full-cell format, this system showed a reversible capacity of 143 mA h g^{-1} in the potential windows of 1.5–2.9 V and exhibited a capacity retention of ~80% at the end of 300 charge/discharge cycles.[48] The Ti^{4+} is an electrochemically inactive ion, but it is one of the most promising dopants to enhance the cycling performance of oxide-based cathode materials.[49,50] The O3-type NaTi$_{0.5}$Ni$_{0.5}$O$_2$ exhibited a specific discharge capacity of 121 mA h g^{-1} (0.2C) with superior cycling performance and excellent rate capability.[51] The Mn-excess P2-type Na$_{2/3}$Mn$_{0.8}$Fe$_{0.1}$Ti$_{0.1}$O$_2$ material showed a discharge specific capacity of ~144 mA h g^{-1} (potential window: 4.0–2.0 V, C-rate: C/10) and retained more than 95% of its initial capacity after 50 cycles of charge/discharge.[52] This system also showed good capacity retention under moisture exposure conditions. Another recently explored high rate capable and stable cathode material is P2-Na$_{0.6}$[Cr$_{0.6}$Ti$_{0.4}$]O$_2$. A symmetric full-cell fabricated by this P2-Na$_{0.6}$[Cr$_{0.6}$Ti$_{0.4}$]O$_2$ material demonstrated a high rate capability of 75% of the initial capacity at a 12C rate.[53] Yet another promising full-cell material is P2-type-Na$_{0.66}$Ni$_{0.17}$Co$_{0.17}$Ti$_{0.66}$O$_2$. This showed a high average potential of 3.74 V (vs Na$^+$/Na) as a cathode material and a safe voltage of 0.69 V as an anode. The symmetric full-cells made of this compound showed an excellent cycle life of ~95% and ~76% capacity retention at the end of 100 and 1000 cycles, respectively, with an operating voltage of ~3.10 V. The specific discharge capacity of the same symmetric full-cell at a 2C rate was reported to be 65 mA h g^{-1} at 2C.[54] The main reason for the improved electrochemical stability of Ti^{4+}-based multi-cationic-layered transition metal oxides is the disordered arrangement of Ti^{4+} in the transition metal layers; which can effectively avoid the M^{3+}/M^{4+} charge ordering, and, thus it can prevent Na$^+$ vacancy ordering, leading to an improved cycling stability. Like Ti^{4+}, other similar electrochemically inactive metal ions, such as Sb^{3+}, Mg^{2+}, Ca^{2+}, etc., are also considered to be good options to improve the electrochemical properties because these metal substitutions can enhance the cycling stability by suppressing/preventing the phase separation.[55] For example, P2-type Na$_{0.67}$[Mn$_{1-x}$Mg$_x$]O$_2$ (0 < x < 0.2) compounds have demonstrated a high specific discharge capacity of ~175 mA h g^{-1} with good cycling performance, due to the presence of the Mg content.[56] Na(Ni$_{2/3}$Sb$_{1/3}$)O$_2$ having a layered honeycomb-like structure has shown a good initial capacity of 130 mA h g^{-1}, with a consistent P3 stacking order throughout the desodiation stages.[57] For this material, the phase transitions occurred mainly in two steps [~3.27 V (P3 → P'3 transition) and ~3.64 V (O3 → P3 transition)] during the redox process. In recent years, Li metal ion substitution has attracted considerable attention. This is mainly because for P2-type-layered oxides, the substituted Li can prevent the frequent P2–O2 phase transformation and retains a P2-stacking even up to a high voltage of 4.4 V during the charging process; resulting in enhanced cycling stability.[58] For O3-phase compounds, the Li$^+$ substitution is believed to be beneficial for improving the electrochemical properties; which is mainly because Li$^+$ ions can help in the formation of an O3 phase, hence, it can assist to reduce the O3–P3 phase transformations. For example, the Li substituted Na[Li$_{0.05}$(Ni$_{0.25}$Fe$_{0.25}$Mn$_{0.5}$)$_{0.95}$]O$_2$ compound has exhibited an excellent capacity of ~180 mA h g^{-1} (0.1C) with superior rate and cycling performance as compared to the material without lithium.[59] Another notable, Li substituted compound is P2–O3-type Na$_{0.66}$Li$_{0.18}$Mn$_{0.71}$Ni$_{0.21}$Co$_{0.08}$O$_{2+x}$. This material also demonstrated an excellent specific capacity (200 mA h g^{-1}) along with a high rate capability [34 mA h g^{-1} (1C)], promising cycling performance (84% capacity retention at the end of 50 cycles (0.2C)], and superior energy density (~640 W h kg^{-1}).[60]

4.2.2 Sodium-Free Metal Oxides

In addition to the widely studied sodium transition metal oxides, the original sodium-free metal oxide with a common chemical formula of MO$_x$ (M = V, Mn, Mo, etc.) has been investigated widely due to its facile synthesis technique and reasonably good electronic conductivity. Many studies have shown that the sodium-free MO$_x$ systems can also exhibit relatively good electrochemical performances.[61–63]

4.2.2.1 Vanadium Oxides

Among various sodium-free metal oxides, vanadium oxide holds very satisfactory electrochemical performances, such as reasonable capacity, good crystal structural flexibility, economic, and material abundance.[64,65] Metastable vanadium dioxide VO$_2$ is considered to be a stand out choice for SIBs as compared to other explored vanadium oxides. This is because of its distinct layered crystal structure, built from edge and corner-sharing distorted octahedral VO$_6$, which allows faster diffusion of sodium ions and insertion to the tunnel-structured VO$_2$.[64] These vanadium oxides can hold one sodium atom without interrupting the crystal structure. When the inserted Na exceeds one, however, it causes a large volume expansion resulting in broken V–O tunnels. In general, for the redox process; during initial discharge process, one sodium ion gets inserted in a VO_2 structure to form $NaVO_2$, and then, in the subsequent charging process, $NaVO_2$ gets oxidized to Na_xVO_2. This single sodium insertion redox process leads to a theoretical specific capacity of 323 mA h g^{-1}. These materials showed an experimental initial discharge capacity of 214 mA h g^{-1} at a current density of 50 mA g^{-1} in the potential window of 1.5–4.0 V (vs Na$^+$/Na). But this material suffered from poor cycling performance due to the irreversible capacity loss and excess sodium ion insertion during the initial cycle. Many studies have demonstrated that nanostructuring is an effective way to mitigate this issue and hence improve the electrochemical performances of VO_2 cathodes. This is because the nanoscaling can introduce more pathways for Na$^+$ transition, to facilitate better access of Na$^+$ between the electrolyte and the electrode, by ensuring the structural stability and superior surface conductivity.[66] Another highly explored vanadium oxide is V_2O_5, mainly because of its superior thermal and chemical stabilities. Studies have shown that these vanadium pentoxides can host sodium insertion/extraction processes.[67] There are two types of V_2O_5 crystal structures, namely, orthorhombic and bilayered. Earlier studies reported that bilayered V_2O_5 with a larger d spacing than orthorhombic V_2O_5 can achieve a good electrochemical performance as a cathode in SIBs. To improve the performances of V_2O_5, scientists have started different approaches, such as nanostructuring the material, introducing secondary phases like conductive carbon, fabricating specific crystal structure with desired exposed facets, etc.[12,62,65] For instance, V_2O_5 materials with large (001) d spacing of ~11.53 Å can easily house sodium ion insertion/extraction processes resulting in better capacity.[62] One of the major drawbacks of orthorhombic V_2O_5 is the slow sodium ion diffusion within the closely packed crystal structure. The crystal structure and the Na ion intercalation/de-intercalation channels along with charge-discharge processes are provided in Figure 4.2b.[65] However, a properly designed nanostructured-surface could mitigate the limitation of this compact structure. A study revealed that the orthorhombic V_2O_5 confined in the nanoporous carbon can have a higher specific capacity of 276 mA h g^{-1} at a current density of 40 mA g^{-1}, as compared to the pristine V_2O_5 (170 mA h g^{-1}).[68] Studies have also demonstrated that the hollow architectures and desired exposed facets are also some of the efficient routes to enhance the performances of the cathodes; these approaches can accommodate a larger strain without any pulverization and can provide proper two-dimensional sodium ion diffusion paths for insertion/extraction processes, respectively.[69]

4.2.2.2 Manganese Oxides

Manganese dioxide (MnO_2) has been investigated as a cathode material due to its large open tunnels structure and interstitial spaces for sodium ion storage and transport.[70] The intercalation of sodium in MnO_2 structures is expected to happen in three sodium insertion steps at three different potentials of 3.34, 2.84, and 2.19 V (vs Na$^+$/Na).[71] The low migration barrier (~0.48 eV) of MnO_2 is also considered to be a positive attribute for faster diffusion of sodium, resulting in rapid charge and discharge leading to high rate performance. The β-MnO₂ nanorods with exposed [111] crystal planes have more than twice (1 × 1) tunnels per unit area than that of α-MnO₂. In a study, these β-MnO₂ have shown an initial specific discharge capacity of 298 mA h g⁻¹ with a promising rate performance.[61] Further, the charge/discharge performance of MnO₂ can be enhanced by fabricating low-dimension, porous structures with properly exposed facets and combined with conductive secondary phases like carbon. Besides, a pre-sodiation technique can be also utilized to enhance the electrochemical performance, especially cycling stability and rate capability; where the pre-intercalated Na between the M–O layers efficiently reduces the irreversible crystal phase transition and thus stabilizes the layered crystal structures.[64]

4.2.2.3 Tunnel-Type Sodium Transition-Metal Oxides

Apart from the layered transition-metal oxides, tunnel-type sodium transition-metal oxides have been also proposed as potential materials for SIBs. The tunnel metal oxide with a common formula of Na$_x$MO$_2$ (Pbam space group) has an orthorhombic crystal structure. In this lattice structure, M^{4+} ions and 1/2 M^{3+} ions will be located in the MO$_6$ octahedral sites, and other ½ M^{3+} ions will be placed in the MO$_5$ square-pyramidal sites.[72] These arrangements lead to the formation of tunnels, which permit the sodium ions to diffuse mainly through the c-direction. In 1971, the most popular tunnel-structured Na$_{0.44}$MnO$_2$ material (theoretical capacity of 121 mA h g⁻¹) with Pbam space group having S-shaped tunnels has been explored.[73] This material was not further investigated during that time, due to the difficulties associated with its hopping mechanism. In recent years, however, many studies have come up with the electrochemical performance analysis of Na$_{0.44}$MnO$_2$;[74–78] due to the facile synthesis route using proven techniques like solid-phase method, hydrothermal reaction, and sol-gel process, etc. The sodium storage properties of this material were first explored in solid polymer electrolyte batteries.[74] Later, the crystal structural variations of Na$_{0.44}$MnO$_2$/C during the charge/discharge process were demonstrated in SIBs using *in-situ* X-ray diffraction (XRD) characteristics.[75] Electrospun, one-dimensional hierarchical Na$_{0.44}$MnO$_2$ material showed a specific discharge capacity of 69.5 mA h g⁻¹ at a high C-rate of 10C in the potential window of 1.5–4 V (vs Na/Na⁺). The high rate capability of this material could be attributed to the high aspect ratio one-dimensional continuous framework and the single crystalline structure with a large S-shaped tunnel.[79] The Na$_{0.44}$MnO$_2$ nanorods prepared using a poly(vinylpyrrolidone)-assisted synthesis technique demonstrated a specific capacity of 122 mA h g⁻¹ in the voltage range of 2–4 V at C/5 rate and showed a high-rate performance of 99 mA h g⁻¹ at 20C with good cycling stability. This study also revealed, by analyzing the soft X-ray spectroscopic study results, that cycling above 3 V could minimize the formation of inactive surface M^{2+}, thus can enhance the cycling stability of the material.[80] In all the aforementioned cases, however, the capacity obtained for $Na_{0.44}MnO_2$ material was very small or the method of preparation was not facile/feasible to make it in bulk quantity; and thus, at this stage, this material is considered to be unsuitable for practical application. The vanadium-based monoclinic $Na_{0.282}V_2O_5$ nanorods prepared using a hydrothermal technique, revealed a 3D tunnel structure along the b-direction. This material showed a specific capacity of 104 mA h g^{-1} at a current density of 0.3 A g^{-1} (1.5–4 V; vs Na/Na$^+$) and with a high cycling stability of 79% over 1000 cycles.[81] To improve the electrochemical properties, several tunnel-structured mixed-cation oxides have been investigated.[72,82] For instance, a binary $Na_{0.66}[Mn_{0.66}Ti_{0.34}]O_2$ system was developed to use as a cathode for full-cell aqueous SIBs (electrolyte: 1 M Na_2SO_4 aqueous solution). This full-cell delivered a promising specific capacity of 76 mA h g^{-1} at 2C rate.[83] The $Na_{0.5}K_{0.1}MnO_2$ nanorods synthesized using a co-precipitation technique have delivered a high specific discharge capacity of 142.3 mA h g^{-1} (0.1C) in the potential window of 1.5–4.3 V (vs Na/Na$^+$) and which maintained a capacity of 94.7 mA h g^{-1} after 100 charge/discharge cycles.[84] Another commonly investigated tunnel-type binary cathode with the Pnma space group is $Na_xFe_xTi_{2-x}O_4$ (x = 1, 0.875).[12] Although, this material has Fe^{3+}/Fe^{4+} redox couple, it could only extract 0.24 Na from the crystal structure even if it was charged up to 4.5 V. With a lower Na content, however, it could extract 0.37 Na from $Na_{0.875}Fe_{0.875}Ti_{1.125}O_4$ with better kinetics, due to the enhanced channel conductivity.[85] These studies revealed that the tunnel-structured material offers improved stability for the Fe^{4+} state as compared to the layered metal oxides. Further, a quaternary $Na_{0.61}[Mn_{0.27}Fe_{0.34}Ti_{0.39}]O_2$ has developed and studies as a cathode. This air-stable system has demonstrated a specific capacity of ~90 mA h g^{-1} (C/10) in the voltage range of 2.6–4.2 V (vs Na/Na$^+$).[86] By considering all the above-mentioned studies, it is anticipated that the appropriate combination of various cations will be an efficient way to improve the electrochemical performances of SIBs.

4.3 Transition-Metal Fluorides

The two main categories of fluorides as cathodes for SIBs are perovskite transition-metal fluorides MF_3 and sodium fluoroperovskites $NaMF_3$ (M = Ni, Fe, Mn). The stronger ionic bonding and electronegativity of fluorine than oxygen made these fluoride-based cathodes a potential candidate for SIBs; especially due to their high operating voltages.

4.3.1 Na-Free Transition-Metal Fluorides

A recent study on amorphous FeF$_3$/C nanocomposite derived from metal-organic frameworks demonstrated a high specific capacity of 302, 146, and 73 mA h g^{-1} at current densities of 15, 150, and 1500 mA g^{-1}, respectively. This system also delivered a promising cycling stability with 0.45% fading per cycle over 100 cycles. These good electrochemical performances of the FeF$_3$/C nanocomposites were attributed to the amorphous structure of FeF$_3$ along with the highly graphited porous carbon framework; which will be a favorable attribute to improve the ionic and electronic transportation and the reaction kinetics of the composite material.[87] Another study revealed a strategy to build a highly insulated FeF$_3$ as high performing cathode materials for SIBs. In this study, a metallic iron and reduced graphene oxide (rGO) conducting matrix was utilized to enhance the conductivity of the FeF$_3$. The prepared FeF$_3$/Fe/rGO composite exhibited a good capacity of 150 mA h g^{-1} (current density: 50 mA g^{-1}), with promising cycle stability and rate capability.[88] Another study investigated a FeO$_{0.7}$F$_{1.3}$/C nanocomposite prepared using a solution technique. This new composite had shown an excellent specific capacity of 496 mA h g^{-1} and a good cycling stability with a minor specific capacity fading of ~0.14% per cycle. Further, this system also exhibited an outstanding energy density of 650 W h kg^{-1}. The charge storage mechanism studies of FeO$_{0.7}$F$_{1.3}$/C revealed that this material exhibits two reaction mechanisms; namely, intercalation and conversion reaction.[89]

4.3.2 Na-Based Transition-Metal Fluorides

Among different NaMF$_3$ materials (M = Mn, Fe, Co, Ni, Cu), NaFeF$_3$ was the only potential compound that could deliver good redox activity. That is, this material can react with Na$^+$ in both insertion and conversion ways.[90] However, some experimental reports and computational simulations suggest the possible electrochemical activity of Mn^{3+}/Mn^{4+} redox couple. The computational analyses revealed that for NaMnF$_3$ materials, the sodium ions could be extracted from the lattice structure at very high voltages of over 5.7 and 4.7 V.[91] But in the present condition with the currently available electrolytes for SIBs, a high potential cannot be accomplished. Thus, with the different perovskite structures, the NaFeF$_3$-based materials will be the only active cathodes with good electrochemical performance for SIBs. For instance, a study on NaFeF$_3$ as a cathode material for SIBs demonstrated a specific discharge capacity of 120 mA h g^{-1} (vs Na/Na$^+$).[92] Another study revealed that the performance of this material is highly dependent on the particles size at higher C-rates. That is, NaFeF$_3$ with smaller particle sizes exhibits better electrochemical performance at C-rates higher than 0.1C.[93] However, this study also demonstrated that this material could deliver specific discharge capacities of 170–180 mA h g^{-1} at 0.01C in the potential window of 1.5 and 4.5 V (vs Na/Na$^+$), despite the NaFeF$_3$ particle size.

Another study utilized a roll-quench processing technique to prepare highly crystalline NaFeF$_3$ materials. This material has delivered an initial specific capacity of ~197 mA h g^{-1} at a current density of 0.076 mA cm^{-2} in the potential range of 1.5 and 4.5 V.[94] This NaFeF$_3$ compound, however, needs to further improve the cycling performance and Coulombic efficiency to meet the practical applications.

4.4 Polyanion Compounds

The success of polyanionic phosphate compounds in LIBs led the widespread investigations of polyanion phosphate and mixed polyanion materials as a cathode system in SIBs. By comparing with monoanionic materials, the redox potential of the transition metal in the polyanion compound can be tuned through an inductive effect by the ionic-covalent characteristic of the bonding (e.g., P–O); and the polyanions (e.g., PO$_4^{2-}$) can shift the metal redox couples (e.g., Fe^{3+}/Fe^{4+}), thus can exhibit a higher operating potential. These polyanionic systems are considered to be very promising candidates for the practical SIBs due to their different properties, such as good structural stability, diversity, superior operating potential, thermal safety, high cycling life, and strong inductive phenomenon resulting from the highly electronegative anions.

4.4.1 Phosphates

NASICON- and olivine-based phosphates are the two widely investigated cathodes for SIBs. Besides these materials, however, there are many other available potential phosphate-based cathode systems for SIBs, such as pyrophosphates, alluaudites, mixed polyanionic type, fluorophosphates, etc.

4.4.1.1 Olivine

The high success of a LiFePO$_4$ cathode in commercial LIBs directed the exploration of the sodium analogue (NaFePO$_4$) as a cathode material for SIBs.[95–97] Among different NaFePO$_4$, the olivine-structured material demonstrates many superior characteristics, such as high theoretical capacity (154 mA h g^{-1}), good operating potential [2.9 V (vs Na$^+$/Na)], etc., for SIBs applications. In general, the olivine NaFePO$_4$ was synthesized through electrochemical Na insertion into hetero-sites of delithiated olivine-structured LiFePO$_4$.[98] Unlike the Li counterpart, during each charging process, the NaFePO$_4$ shows a transitional phase of Na$_{0.7}$FePO$_4$ at a voltage of 2.95 V.[99–101] The NaFePO$_4$ compound fabricated by common synthesis technique, however, is thermodynamically stable, but electrochemically inactive.[102–105]

Besides, the structure of the material fabricated through a normal route will not be olivine, but maricite, wherein Na$^+$ and Fe^{2+} are positioned opposite as compared to olivine-structured LiFePO$_4$. The structural framework of an olivine-NaFePO$_4$ compound is highly favorable for sodium ion insertion and transmission (Figure 4.3a),[106] leading to high reversibility and good cycling stability. However, this system suffers from poor electronic conductivity and low sodium ion diffusion coefficient, thus leading to a poor practical capacity even though the material has a high theoretical capacity. These drawbacks can be avoided by decreasing the particle size and/or coating the particle with appropriate secondary phases like conducting carbon.[107,108] Apart from the usual carbon coating, a conductive-polymer wrapping with a high mechanical pliability is also an excellent alternative to improve the overall electrochemical performances. A study on olivine NaFePO$_4$/polythiophene cathode system reported that this material could deliver a superior specific capacity of 142 mA h g^{-1} at a current density of 10 mA g^{-1} [2.2–4.0 V (vs Na/Na$^+$)] and could also demonstrate a high capacity retention of 94% after 100

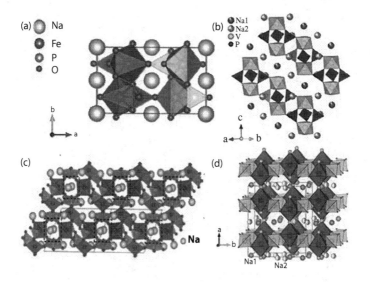

FIGURE 4.3
(a) Crystal structure of NaFePO$_4$ (Reprinted from *J. Power Sources*, 291, Fernández-Ropero, A. J. et al., Electrochemical characterization of NaFePO$_4$ as positive electrode in aqueous sodium-ion batteries, 40, Copyright 2015, with permission from Elsevier), (b) crystal structure Na$_3$V$_2$(PO$_4$)$_3$ (Reprinted from *Electrochem. Commun.*, 14, Jian, Z. et al., Carbon coated Na$_3$V$_2$(PO$_4$)$_3$ as novel electrode material for sodium ion batteries, 86, Copyright 2012, with permission from Elsevier), (c) schematic illustration of the crystal structure (viewed from bc plane), showing tunnels favoring Na$^+$-ion migration [highlighted by dotted rectangles] (Reprinted from *Electrochem. Commun.*, 24, Barpanda, P. et al., Sodium iron pyrophosphate: A novel 3.0 V iron-based cathode for sodium-ion batteries, 116, Copyright 2012, with permission from Elsevier), and (d) crystal structure of Na$_3$V$_2$(PO$_4$)$_2$F$_3$ (Reprinted with permission from Bianchini, M. et al., *Chem. Mater.*, 24, 4238, 2014. Copyright 2014 American Chemical Society).

charge/discharge cycles.[109] As mentioned above, it should be noted that it is very difficult to prepare the olivine-structured NaFePO$_4$ by the usual high-temperature route, because this route could lead to the formation of a thermodynamically stable, but electrochemically inactive maricite-structured NaFePO$_4$.[110] A recent study, however, demonstrated that 50 nm-sized maricite-structured Na$_{1-x}$FePO$_4$ material could deliver a reversible specific capacity of 142 mA h g^{-1} in the potential window of 1.5–4.5 V (vs Na$^+$/Na), with a good cycling stability of 95% capacity retention at the end of 200 charge/discharge cycles. This superior electrochemical performance of maricite-structured Na$_{1-x}$FePO$_4$ could be attributed to the enhanced Na mobility due to the phase transformation from maricite to amorphous.[111] Amorphous-type FePO$_4$ has been also investigated as a cathode material for SIBs, owing to the fact that it can be processed easily through low temperature routes.[112] The single-phase insertion/extraction process of amorphous FePO$_4$ can be beneficial in SIBs due to its monotonous voltage change. Nevertheless, this material suffers from a number of drawbacks, such as low electronic conductivity, poor diffusion coefficient, and the absence of a potential plateau. To mitigate these drawbacks, many studies have come up with different solutions, such as morphology and dimension tuning, surface coating with secondary conducting carbon phases, etc.[60] For example, an amorphous FePO$_4$ with mesoporous structures demonstrated a high initial specific capacity of 151 mA h g^{-1} at a current density of 20 mA g^{-1} (1.5–4.0 V; vs Na$^+$/Na), with ~94% capacity retention over 160 cycles of charging/discharging. These improved electrochemical performances could be attributed to the enhanced sodium ion transportation and reaction kinetics.[113] The conducting carbon coating is an efficient method, which not only improves the electronic conductivity, but also enhances the cycling stability. The highest capacity reported for amorphous FePO$_4$ was 166 mA h g^{-1} at C/10 rate and which was prepared through a virus-templated technique. The prepared material was an amalgamation of bio-templated FePO$_4$ and single-wall carbon nanotubes (SWCNTs).[101] Another way to enhance the electrochemical performances of materials is to create a higher dimensional pathway for sodium ion transportation and diffusion. In the case of most NaFePO$_4$ cathodes, one of the main reasons not to achieve a practical capacity close to the theoretical capacity is its restricted sodium ion diffusion and transportation, due to the one-dimensional channel. The above issue could be address by partial substitution of fluorine-atom. For instance, the fluorine-doped NaFePO$_4$ demonstrated improved electrochemical properties, due to its layered crystal structure with strong and stable 2D sodium ion channels.[114,115] The other explored olivine structures, as cathode materials for SIBs, are NaMn$_x$M$_{1-x}$PO$_4$ (M = Fe, Ca, Mg) compounds.[107] These materials, however, demonstrated sloping charge/discharge characteristics. Also, by increasing the Mn concentration in a NaMn$_x$Fe$_{1-x}$PO$_4$ system demonstrates reduced electrochemical activity. This phenomenon is just similar to the Li analogue (LiMn$_{1-x}$Fe$_x$PO$_4$), which shows the unfavorable electrochemical activity of Mn in phosphate materials.[116]

4.4.1.2 Sodium Super Ionic Conductor (NASICON)

NASICON structures with the common chemical formula $Na_xM_2(PO_4)_3$ (M: different transition metals) were originally investigated to use as solid electrolytes for batteries due to their high sodium ion conductivity because of the open three-dimensional crystal structure.[117,118] A commonly explored NASICON compound is $Na_3V_2(PO_4)_3$.[119] This three-dimensional open crystal framework of $Na_3V_2(PO_4)_3$ is very advantageous in terms of electrochemical performance as compared to the low-dimensional Na^+ pathways in $NaFePO_4$ (Figure 4.3b).[120] The three-dimensional crystal structure of $Na_3V_2(PO_4)_3$ is created by corner-sharing tetrahedral PO_4 and octahedral VO_6 units, where the units are gathered in a three-dimensional manner to form $V_2P_3O_{12}$, this structure has large tunnels with rhombohedral interstitial sites which can fully/partially house sodium ions by a perturbation-free lattice insertion/extraction process. These large tunnels in the NASICON structure make sure of the fast sodium ion diffusion in a three-dimensional manner. The main attractions of this material are: (1) superior ionic mobility; (2) promising theoretical specific capacity (117.6 mAh g^{-1}), when it is cycled between 2.4 and 3.7 V, wherein two sodium ions can participate in the redox process; (3) two flat/stable plateaus at 3.4 and 1.6 V, corresponds to the redox couples V^{4+}/V^{3+} and V^{3+}/V^{2+}, respectively, thus can be used as both cathode and anode; (4) infinitesimally small hysteresis voltages, ~0.1 V (V^{4+}/V^{3+}) and 0.06 V (V^{3+}/V^{2+}); and (5) good charged state thermal stability (~450°C), validating its superior safety for practical applications.[121] In 2002, sodium storage properties of $Na_3V_2(PO_4)_3$ were first explored for the redox couple V^{4+}/V^{3+} and V^{3+}/V^{2+} when the sample cycled at 1.2–3.5 V, the electrode material showed a high specific capacity of 140 mAhg^{-1}, but with poor cycling stability.[122] Later, a symmetric full-cell fabricated out of $Na_3V_2(PO_4)_3$ demonstrated a low operating potential (~1.6 V), however, a poor cycling stability.[123] All the studies on pristine $Na_3V_2(PO_4)_3$ demonstrated poor cycling and rate performances. The reason for this poor performance of this NASICON-structured $Na_3V_2(PO_4)_3$ compound is its inherent low electrical conductivity (~10^{-9} S cm^{-1}) due to the slightly distorted arrangement of octahedral VO_6, hence leading to the low cycling stability and rate performance, which restricts its practical battery performance and further commercialization.[123] Globally, scientists have tried different routes to overcome this issue, which include nanostructuring the material, wrapping the material with a thin layer of conducting carbon or any other conducting material, doping with secondary cations, etc., which could facilitate redox kinetics and could achieve higher conductivity. From a profitable and bulk production aspect, however, conducting carbon coating is undoubtedly an efficient method to improving the electrochemical properties.[125–127] For instance, the porous $Na_3V_2(PO_4)_3$/C composite prepared using the sol-gel technique followed by lyophilization and annealing leads to a specific discharge capacity of ~118 mAhg^{-1} at 0.05C and 92.7% initial capacity retention over 50 when cycled at 2.7–4 V (vs Na^+/Na).[125] A honeycomb-structured $Na_3V_2(PO_4)_3$/C showed 93.6% of its initial capacity at the end of 200 cycles of charge/discharge

(1C rate).[128] Another study utilized a solution-based template route and reported a specific discharge capacity of ~113 mAh·g^{-1} and a capacity retention of 86.7% over 1000 cycles (1C) and 50% capacity retention after 30,000 cycles at 40C.[129] The in-situ grown rGO/Na$_3$V$_2$(PO$_4$)$_3$ compound prepared using a microwave solvothermal technique exhibited more than 85% capacity retention after 6000 charge/discharge cycles even at 10C. This study also demonstrated deep discharges by extending the voltage range to 1.3–3.8 V (vs Na/Na$^+$), leading to a high specific discharge capacity of ~165.4 mAh g^{-1} (1C).[121] The secondary cation substitution is also an effective way to improve the electrochemical performances of Na$_3$V$_2$(PO$_4$)$_3$ materials. The Mg-doped Na$_3$V$_{2-x}$Mg$_x$(PO$_4$)$_3$/C composites exhibited promising electrochemical performances. Among different Mg-doped Na$_3$V$_{2-x}$Mg$_x$(PO$_4$)$_3$/C, Na$_3$V$_{1.95}$Mg$_{0.05}$(PO$_4$)$_3$/C demonstrated the superior rate capability of 112.5 mA h g^{-1} (1C) and 94.2 mA h g^{-1} (30C) with a cycle performance of 81% initial capacity retention after 50 cycles at 20C. This improved cycling and rate performance were attributed to the enhanced structural stability and improved ionic/electronic conductivity through Mg doping.[130] Another study examined the effect of Mn doping on the electrochemical performance, where the Na$_3$V$_{2-x}$Mn$_x$(PO$_4$)$_3$/C (0 < x < 0.7) was prepared using a sol-gel process. The optimized sample [Na$_3$V$_{1.7}$Mn$_{0.3}$(PO$_4$)$_3$/C] showed a capacity of 104 mA h g^{-1} at 0.5C in the potential window of 2.0 and 4.3 V (vs Na/Na$^+$), which was higher than that of the pristine sample.[131] By considering all the aforementioned studies, this NASICON-structured Na$_3$V$_2$(PO$_4$)$_3$ material demonstrates promising structural and thermal stability, as well as good electrochemical performances. Thus, this material is considered to be a possible candidate for future SIBs applications. Apart from the widely explored Na$_3$V$_2$(PO$_4$)$_3$ material, there are many other NASICON-structured materials explored from earth-abundant materials such as Fe, Mn, Ni, etc., for SIBs. The iron-based NASICON-type Na$_3$Fe$_2$(PO$_4$)$_3$ materials have been rarely studied, however, mainly due to their low voltage plateau (~2.5 V) and the poor specific capacity associated with the Fe^{3+}/Fe^{2+} redox couple. Also, most of the studies on iron-based NASICON materials have been limited to the Fe^{3+}/Fe^{2+} redox couple and did not investigate the potential high voltage Fe^{3+}/Fe^{4+} redox couple, due to the complexity in achieving a stable Fe^{4+} state. A recent study, however, investigated the reversible nature of the Fe^{3+}/Fe^{4+} redox couple and the electrochemical properties of Na$_3$Fe$_2$(PO$_4$)$_3$ materials prepared using a solid-state technique. The NASICON-structured Na$_3$Fe$_2$(PO$_4$)$_3$ wrapped in the conducting carbon, demonstrated a specific discharge capacity of ~109 mA h g^{-1} with a cycling performance of greater than 96% initial capacity retention over 200 charge/discharge cycles. Moreover, the study also demonstrated the reversible nature of the Fe^{3+}/Fe^{4+} redox couple by analyzing the self-discharge behavior, the cycling stability characteristics, and the Coulombic efficiency of the prepared electrode.[132] Apart from their promising electrochemical properties, NASICON-based compounds have also shown good compatibility with aqueous electrolytes. This fact has led more scientists across the world to explore new superior redox active NASICON systems.

4.4.1.3 Vanadyl Phosphate

The vanadyl phosphate NaVOPO$_4$ crystal structure is formed from chains of corner-sharing distorted octahedral VO$_6$ units along the c-axis, which are connected together by tetrahedral PO$_4$ units. The theoretical capacity and energy density of vanadyl composites are 165 mA h g^{-1} and 490 W h kg^{-1}, respectively, with an operating potential of 3.4–3.5 V and V^{4+}/V^{5+} redox couple. These theoretical values are higher than many other phosphate systems, mainly because of the lighter nature of the vanadyl compounds. The first explored monoclinic-structured NaVOPO$_4$ compound for SIBs delivered a capacity of 90 mA h g^{-1} (C-rate: C/15) with an average potential of 3.6 V.[133] The β-NaVOPO$_4$, analogous to β-LiVOPO$_4$, having an orthorhombic symmetry is generally affected with the protons in the crystal structure; which leads to large irreversible capacity. Hence, this system could only achieve much lower practical capacity [for example: 60 mA h g^{-1} (C/20) in the voltage range of 4.3–1.5 V] than its theoretical capacity.[134] Among the different polymorphs of VOPO$_4$, α-NaVOPO$_4$ shows tetragonal symmetry and demonstrates a layered crystal structure, where the large inter-planes of VOPO$_4$ are beneficial to easily house sodium ions and aid its fast diffusion. A recent work on α-NaVOPO$_4$ compound reported that these phosphate systems along with rGO sheets can demonstrate high capacities of 110 mA h g^{-1} (0.67 Na$^+$) and 150 mA h g^{-1} (0.90 Na$^+$).[135]

4.4.1.4 Pyrophosphates

As compared to the normal phosphate systems, the pyrophosphate (P$_2$O$_7$) anion is considered to be thermodynamically more stable, thus can serve as a potential cathode for SIBs.[136] Moreover, pyrophosphate with a common chemical formula of Na$_2$MP$_2$O$_7$ (M = Fe, Mn, Co) can be easily obtained through the thermal decomposition of phosphates with oxygen evolution.[136,137] The robust and stable 3D crystal framework of the pyrophosphate anions can facilitate the sodium ion accommodation; also the crystal structure configuration of Na$_2$MP$_2$O$_7$ could provide various opportunities to host the sodium ion insertion/extraction process in the voltage window of 2.0–4.9 V (Figure 4.3c).[138–140] The Na$_2$CoP$_2$O$_7$ compound could exist in three different polymorphs, which include orthorhombic, tetragonal, and triclinic crystal phases. Both the orthorhombic and tetragonal polymorphs of Na$_2$CoP$_2$O$_7$ are shown to have layered crystal structure, with layers of [Co(P$_2$O$_7$)]$^{2-}$ constructed from oxygen-linked tetrahedral CoO$_4$ units and their adjacent four P$_2$O$_7$ units. This unique layered crystal structure presents continuous channels for sodium transportation leading to an efficient sodium ion intercalation/deintercalation process.[138] The triclinic polymorph phase has also shown redox activity; however, the thermodynamic instability restricts its employment as a practical cathode material for SIBs. Nonetheless, a recent study successfully synthesized a stable triclinic polymorph phase

($Na_{2-x}CoP_2O_7$) using a sodium deficiencies-stabilization strategy. The as prepared material could deliver a capacity of ~80 mA h g^{-1} (C/20) with three different potential plateaus of 3.95 V, 4.33 V, and 4.43 (vs Na/Na$^+$). This system also demonstrated a superior energy density of 344 W h kg^{-1} compared to the orthorhombic polymorph phase (240 W h kg^{-1}).[141] As compared to $Na_2CoP_2O_7$, the triclinic polymorph phase of $Na_2FeP_2O_7$ has exhibited better thermal stability of up to 600°C without any decomposition or oxygen evolution, thus has been considered to be a potential and safe cathode material for practical and bulk-scale SIBs applications.[139] A study on triclinic $Na_2FeP_2O_7$ crystal structure with a well-defined channel reported a reversible specific capacity of ~90 mA h g^{-1} in the potential range of 2.0–4.5 V with a good cycling stability.[142] Further, the $Na_2FeP_2O_7$ compound with a combination of conducting carbon phases has shown to enhance the rate capability. For instance, a $Na_2FeP_2O_7$/CNTs composite could deliver a specific discharge capacity of 86 mA h g^{-1} (2.3–4.0 V; vs Na$^+$/Na) after 140 cycles of charge/discharge at 1C and 68 mA h g^{-1} at 10C.[143] Recently, an investigated off-stoichiometric $Na_{3.12}Fe_{2.44}(P_2O_7)_2$ compound exhibited a theoretical capacity of 117.4 mA h g^{-1} when 2.44 sodium was being reversibly deintercalated/intercalated in the structure.[144] Later, $Na_{3.12}Fe_{2.44}(P_2O_7)_2$/multi-wall carbon nanotube (MWCNT) composite cathode was developed and showed a specific discharge capacity of >100 mA h g^{-1} (0.15C) with a promising cycling life over 50 cycles.[145] Another potential cathode system for SIBs is triclinic $Na_2MnP_2O_7$. A study on $Na_2MnP_2O_7$ demonstrated a specific discharge capacity of ~80 mA h g^{-1} at an average voltage of 3.6 V.[146] It is speculated that the triclinic $Na_2MnP_2O_7$ polymorph can withstand huge structural variations such as Jahn-Teller distortions due to its crystal structure arrangements.[146] In most studies, however, $Na_2MnP_2O_7$ pyrophosphates demonstrated sloping charge/discharge characteristics,[146,148] restricting its practical application in full-cell SIBs. Apart from the aforementioned compounds, many other mixed-cation pyrophosphates have been also proposed to improve the electrochemical properties.[149,150]

4.4.1.5 Fluorophosphates

The search for potential cathode materials for SIBs led to the invention of a new class of compounds by integrating F$^-$ ions with phosphates.[151,152] These fluorophosphates have demonstrated higher operating voltage than phosphates, due to the strong inductive nature of the fluorine, which can easily tune the electrochemical properties of the M^{3+}/M^{4+} redox couples (M = Ti, Fe, V) present in the material.[153,154] Among the different fluorophosphates, vanadium-based materials have attracted considerable attention due to the presence of the highly redox active V^{3+}/V^{4+} couple, with high operating potential (~3.9 V).[155] These fluorophosphates can be categorized into three different polymorphs, namely, tetragonal $NaVPO_4F$, $Na_3(VO)_2(PO_4)_2F$, and $Na_3V_2(PO_4)_2F_3$. Recent researches have revealed that these compounds can demonstrate high electrochemical performances, similar to that of conventional electrode materials for LIBs. For instance, a recent study on NaVPO$_4$F material has demonstrated a high capacity of 133 mA h g^{-1} (0.1C) with a flat potential discharge plateau at ~3.33 V and a superior cycling stability of ~82% over 2500 cycles (1C). This cathode also has shown a good long-term and high-rate cycling performance of 81% and 77% capacity retention of over 10,000 cycles at 10 and 20C.[156] A thermally stable NASICON-structured Na$_3$V$_2$(PO$_4$)$_2$F$_3$ can demonstrate a high theoretical capacity of 128 mA h g^{-1}, corresponding to the V^{3+}/V^{4+} redox couple, with two sodium ions extraction per formula unit. A study on Na$_3$V$_2$(PO$_4$)$_2$F$_3$ material revealed a specific capacity of 120 mA h g^{-1} with two potential plateaus at 4.2 and 3.7.[157] Another study on NASICON-structured Na$_3$V$_2$(PO$_4$)$_2$F$_3$ prepared via carbothermal reduction reaction has shown to deliver a specific discharge capacity of ~111.6 mA h g^{-1} with a good cycling life with a capacity retention of 97.6% after 50 cycles.[158] The crystal structure of Na$_3$V$_2$(PO$_4$)$_2$F$_3$ is provided in Figure 4.3d. In the crystal structure, V is located in the center of the octahedral VO$_4$F$_2$, forming V$_2$O$_8$F$_3$ bioctahedral units, which are alternately connected by tetrahedra PO$_4$. This forms a stable three-dimensional structure with large tunnels in the [110] and [1$\bar{1}$0] directions, resulting in better Na ion mobility.[159] Among all the other vanadium-based sodium fluorophosphates with the V^{4+}/V^{3+} redox couple, such as NaVPO$_4$F, NaV$_{1-x}$Cr$_x$PO$_4$F, and Na$_{1.5}$VOPO$_4$F$_{0.5}$; the Na$_3$V$_2$(PO$_4$)$_2$F$_3$ exhibited the highest average voltage (~3.95 V) and specific capacity.[157,160-162] A computational study on NASICON-structured Na$_3$GaV(PO$_4$)$_2$F$_3$ revealed that the performance limiting factors in these structures are not basically redox-limited, but are site-limited.[163] This understanding could be beneficial while designing high performing fluorophosphate materials for SIBs applications. Another material with a chemical formula of Na$_3$V$_2$O$_{2x}$(PO$_4$)$_2$F$_{3-2x}$ ($0 \leq x \leq 1$) could deliver promising discharge capacities with two distinct potential plateaus at 3.6 V and 4.1 V.[164,165] However, the electrochemical activity of the Na$_3$V$_2$O$_2$(PO$_4$)$_2$F prepared through a hydrothermal technique always demonstrated micro-sized particles with poor conductivity.[166] A study on Na$_3$V$_2$O$_2$(PO$_4$)$_2$F nanowires wrapped with conductive RuO$_2$ reported a specific discharge capacity of 120 mA h g^{-1} and 95 mA h g^{-1} at the C-rates of 0.1C and 20C, respectively, with over 1000 cycles stability.[167] The promising electrochemical performances could be attributed to the presence of conductive and uniform layers of RuO$_2$ on the Na$_3$V$_2$O$_2$(PO$_4$)$_2$F nanowires. Recent studies have demonstrated that a V^{3+}/V^{4+} mixed valence material can enhance the electrochemical properties because of their inherent high conductivity.[165,168] For instance, Na$_3$V$_2$O$_{2x}$(PO$_4$)$_2$F$_{3-2x}$ has shown high operating voltages of 3.6 and 4.1 V, promising capacity, and good cycling performances.[165] A recent study on Na$_3$V$_2$O$_{2x}$(PO$_4$)$_2$F$_{3-2}$/MWCNTs material prepared through an aqueous route has demonstrated a cycling performance of over 1100 cycles. Moreover, the full-cell fabricated using Na$_3$V$_2$O$_{2x}$(PO$_4$)$_2$F$_{3-2}$/MWCNTs as cathode material has presented an energy density of about 84 W h kg^{-1} along with a working potential of 1.7 V.[168] Another study revealed that the oxidation state of vanadium in Na$_3$V$_2$O$_{2x}$(PO$_4$)$_2$F$_{3-2}$ is +3.8. Further, an in-depth structural analysis of this material displayed that the V^{4+}/V^{5+} redox couple is active when the material is cycled between 2.8 and 4.3 V, whereas V^{3+} remains unchanged.[169] A newly explored $Na_3(VO_{1-x}PO_4)_2F_{1+2x}$ (0 < x < 1) compound, prepared through a low temperature solvothermal technique, demonstrated a promising specific discharge capacity of 112 mA h g^{-1} (C/5) with an average voltage of 3.75 V and good rate (73 mA h g^{-1} at a 10C), and cycling performances (90% retention after 1200 cycles at 2C).[170] Moreover, it has been reported that partial substitution of oxygen is a promising method to create a multi-electron transfer of redox reaction in vanadium. For example, the $Na_3(VO_{1-x}PO_4)_2F_{1+2x}$ compound delivered discharge specific capacities of 120–130 mA h g^{-1} with an average potential of 3.8–3.9 V (vs Na$^+$/Na), having a multi-electron transfer mechanism. By modifying the polyanion group with the $V^{3.8+}/V^{5+}$ redox couple, the $Na_3(VO_{0.8})_2(PO_4)_2F_{1.4}$ material has shown to undergo a redox reaction with 1.2-electron transfer, that is an extra 0.2 electrons per formula unit, which leads to a high theoretical specific energy of ~600 W h kg^{-1}.[171] Another compound with sodium rich chemistry [$Na_4V_2(PO_4)_2F_3$] has also demonstrated the multiple electron mechanism.[172] Moreover, the full-cell fully prepared by utilizing the sodium-excess $Na_4V_2(PO_4)_2F_3$ compound as the cathode demonstrated the advantages of using this material as reservoirs to compensate the sodium losses during solid electrolyte interphase (SEI) layer formation, leading to more than a 10% increment in energy density. Other commonly investigated fluorophosphates materials for SIBs are Na_2MPO_4F (M = Fe, Mn, Co, Ni) and $Na_2Fe_xM_{1-x}PO_4F$.[173] A study on Na_2FePO_4F demonstrated that the utilization of conducting carbon coating can enhance the rate and cycling performance of this material as compared to the pristine system.[154] The pure the Na_2MnPO_4F ($P2_1/n$) materials have low inherent electrochemical activity, however, a study revealed that the nanostructuring of the Na_2MnPO_4F compound can improve its electrochemical activity, leading to a promising specific capacity of 120 mA h g^{-1} within the voltage window of 1–4.8 V, but with a short cycle life.[174] Although most of these fluorophosphate materials show good electrochemical storage properties, the chance of fluorine gas formation during bulk scale production causes some environmental concerns, and needs to be taken care.

4.4.1.6 Mixed Polyanion

The amalgamation of various polyanion units could be a promising way to enhance the electrochemical properties of SIBs by creating new open crystal/tunnel structures with an adjustable redox couple and unique multi-dimensional pathways for improved ionic conductivity. To illustrate, $Li_xNa_{4-x}Fe_3(PO_4)_2(P_2O_7)$ (x = 0–3) polyanionic material could achieve a specific discharge capacity of 129 mA h g^{-1} with a specific energy density of 380 W h kg^{-1} in the average operating voltage of 3.2 V (vs Na$^+$/Na).[175] The computational analysis demonstrates that the collective (PO$_4$) and (P$_2$O$_7$) polyanionic groups can offer a 3D crystal structure to house the abundant Fe redox-active center and sodium ions, thus leading to minimal

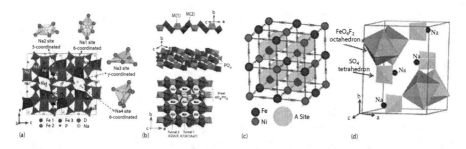

FIGURE 4.4
(a) Schematic illustration of Na$_4$Fe$_3$(PO$_4$)$_2$(P$_2$O$_7$) and order of sodium extraction from the structure of Na$_x$Fe$_3$(PO$_4$)$_2$(P$_2$O$_7$) (1 ≤ x ≤ 4) [Reprinted with permission from *Chem. Mater.* 2013, 25, 3614. Copyright (2013) American Chemical Society], (b) figure (top) view of the chain showing the distorted octahedral M(1) and M(2) sites. (middle) View of the sheet constituted by octahedral MO6 and tetrahedral PO$_4$. (bottom) View of the alluaudite structure in the ab plane [Reprinted with permission from *Chem. Mater.*, 2010, 22, 19, 5554–5562. Copyright (2010) American Chemical Society], (c) Prussian blue face-centered cubic crystal structure, where the Fe and Ni are bound by bridging CN ligands [Reprinted with permission from *Nano Lett.*, 2011, 11, 5421. Copyright (2011) American Chemical Society], and (d) crystal structure of NaFeSO$_4$F [Reprinted with permission from *Chem. Mater.*, 2011, 23, 2278. Copyright (2011) American Chemical Society].

volume changes and good cycling stability (Figure 4.4a).[176,177] A study on [Na$_4$Fe$_3$(PO$_4$)$_2$P$_2$O$_7$] with (PO$_4$) and (P$_2$O$_7$) mixed polyanion group reported a capacity of ~ 110 mA h g^{-1} at C/40.[175] The average operating potential for the Fe^{3+}/Fe^{2+} redox couple in the combined system showed higher than that for the individual phosphates and pyrophosphates. There are many similar potential polyanionic materials to be explored further with a general chemical formula Na$_4$M$_3$(PO$_4$)$_2$P$_2$O$_7$ (M = Co, Ni, Mn, Mg, etc.).[178] For instance, Mn-based Na$_4$Mn$_3$(PO$_4$)$_2$(P$_2$O$_7$) material showed a superior energy density of 416 W h kg^{-1} with an average potential of 3.8 V when it is employed as the cathode material in SIBs.[179] The crystal structure of Na$_4$Mn$_3$(PO$_4$)$_2$(P$_2$O$_7$) offers 3D sodium ion diffusion pathways with low activation barriers to facilitate fast sodium ion insertion/extraction processes, and notably undergoes an anomalous Jahn-Teller effect, where the sodium ion mobility is getting enhanced along with Jahn-Teller distortions (Mn^{3+}) rather than getting reduced (as observed for most electrode materials). Further, many studies have reported that the nanostructuring and conducting carbon coating are effective strategies to enhance the electrochemical properties of the mixed polyanion materials. For example, a 3.8 V Na$_7$V$_4$(P$_2$O$_7$)$_4$(PO$_4$) nanorod with V^{3+}/V$^{3.5+}$/V^{4+} redox reaction can deliver a specific capacity (~92.1 mA h g^{-1}) which is close to its theoretical specific capacity (92.8 mA h g^{-1}) at 0.05C with 95% cycling stability after 200 charge/discharge cycles.[180] For this compound, it is believed that the 1D nanostructure facilitates its reversible sodium intercalation/deintercalation processes, thus leading to the enhanced rate performance and cycling life. Recent studies on a carbonophosphate with a common chemical formula (Na$_3$MPO$_4$CO$_3$) (M = Fe, Co, Mn) demonstrated a two-electron intercalation process, where both the M^{2+}/M^{3+} and the M^{3+}/M^{4+} redox couples are electrochemically active, displaying a great potential as a cathode material in large-scale SIBs application owing to its superior energy density.[181] The sidorenkite Na$_3$MnPO$_4$CO$_3$ is another potential cathode material for SIBs. This material could demonstrate a promising discharge specific capacity of ~125 mA h g^{-1} with an average voltage of 3.7 V, and a specific energy density of 374 W h kg^{-1} because of the presence of two activate redox couples (Mn^{2+}/Mn^{3+} and Mn^{3+}/Mn^{4}).[182]

4.4.1.7 Alluaudites

The alluaudite materials exhibit crystal structures made from both polyhedral (MO$_n$) and tetrahedral (PO$_4$)$^{3-}$ units with a common chemical formula of X^{1}X^{2}M^{1}M$_2^2$(PO$_4$)$_3$, where X^1 and X^2 are cations occupied at different sites in the tunnel structures aligned along the c-axis, which is created by chains of edge-sharing MO$_6$ octahedra units coupled with PO$_4$ tetrahedra units (Figure 4.4b).[110,183,184] Initially, these compounds were tested as a Li insertion host.[110] The alluaudite materials with chemical formulas NaFe$_3$(PO$_4$)$_3$, NaMnFe$_2$(PO$_4$)$_3$, and Li$_x$Na$_{2-x}$FeMn$_2$(PO$_4$)$_3$ were explored, but the cycling performance was observed to be very poor with high voltage polarization.[110,151] Later studies have come up with the solution to enhance the cycling performances of these compounds by decreasing the particle size or coating these compounds with conducting carbon. For instance, alluaudite Na$_{1.702}$Fe$_3$(PO$_4$)$_3$ showed improved specific discharge capacity (140.7 mAh g^{-1}) with high cycling stability after subjecting this material to ball milling and carbon coating.[185] A novel alluaudite Na$_2$VFe$_2$(PO$_4$)$_3$/C composite prepared through a sol-gel route has demonstrated a 3D [VFe$_2$(PO$_3$)$_3$]$^{2-}$ crystal framework with two tunnels to provide fast sodium ion transportation, resulting in enhanced electrochemical properties. When this composite was employed as a cathode material for SIBs, it demonstrated a specific discharge capacity of 65.8 mA h g^{-1} at a current density of 5 mA g^{-1} with a 78% capacity retention after 100 cycles. This study also reported that this compound exhibits multi-electron reactions which include the Fe^{2+}/Fe^{3+}, V^{2+}/V^{3+}, and V^{3+}/V^{4+} redox couples.[186]

4.4.2 Hexacyanoferrates

Prussian blue (PB) with a chemical formula KFeFe(CN)$_6$ and its analogs [Na$_2$M[Fe(CN)$_6$], M = Fe, Co, Mn, Ni, Cu, etc.] are a large entity of transition-metal hexacyanoferrates. These compounds have many superior properties to meet the SIBs applications, such as open crystal frameworks, high theoretical capacity, fast diffusion channels for transportation of sodium ions, abundant redox-active sites, two-electron redox reaction features, and good structural stability.[187] In particular, because of the large diffusion channels and interstices in the lattice structures, PBs are one of the few compounds that can host larger alkali cations, such as sodium and potassium. Moreover, for PBs, each molecular formula could contain two redox active sites $M^{+2/+3}$ and $Fe^{+2/+3}$, indicating that the PBs can reach two-electron redox capacity, corresponding to two sodium ion storage per formula unit (Figure 4.4c).[188] In recent years, many scientific groups have started investigating the sodium analogue of this hexacyanoferrate material for SIBs applications.[188–192] In 2011, a research group investigated the potential electrochemical characteristics of $K_xCuFe(CN)_6$ and $K_xNiFe(CN)_6$ in an aqueous cell, with sodium or potassium ion insertion/extraction.[188,189] Later, another research group evaluated the electrochemical performances of $A_xMFe(CN)_6$ (A = K, Na; M = Ni, Cu, Fe, Mn, Co, and Zn) in an organic electrolyte and have reported good rate and cycling performances.[191,192] Material with a chemical formula of $Na_4Fe(CN)_6$/C exhibited a specific discharge capacity of 90 mA h g^{-1} with good rate capability.[193] A recent study demonstrates that the Na_2FeFe-PB can exhibit a superior specific discharge capacity of ~160 mA h g^{-1} with an average voltage of 3.1 V and a high specific energy density of 496 W h kg^{-1}.[194] The Na_2MnMn-PBA can even reach a much higher specific discharge capacity of 209 mA h g^{-1} at a higher average voltage of 3.50 V (vs Na/Na$^+$) with a very high specific energy density of 730 W h kg^{-1}.[195] These high energy densities demonstrate that the PB materials can even exceed those of commercially available LIB materials, such as spinel-structured $LiMn_2O_4$ (~430 W h kg^{-1}) and olivine-structured $LiFePO_4$ (~530 W h kg^{-1}). Also, apart from the wide selection of M metals in the common chemical formula, the Fe element in the PBs can also be substituted by a wide range of other transition-metals with multiple valence states, such as Co, Ni, Mn, Cu, and Zn, leading to a series of structurally similar, but electrochemically tunable cathode host materials.[189] A recent study has reported that the difficulty to obtain the full discharge specific capacity in PB can be mitigated by doping PB with nickel ions (1%–10%). This study also demonstrates that the nickel doping can improve the electrochemical storage capacity and sodium ion diffusion. In particular, PB doped with 3% nickel ions displayed a discharge specific capacity of 117 mA h g^{-1}, where the 50 mA h g^{-1} was ascribed to the low-spin $Fe^{2+}C_6/Fe^{3+}C_6$ redox couple.[196] Another scientific group designed a $Na_2Ni_xMn_yFe(CN)_6$ (PBMN) compound by substituting electrochemically inert nickel sites for part of the manganese redox active sites. This PBMN compound demonstrated a lower specific discharge capacity of ~118.2 mAh g^{-1} due to the partial substitution of active manganese sites with inactive nickel. However, the PBMN could attain specific capacity retention of 83.8% over 800 cycles. This improved capacity retention could be attributed to the electrochemical inactiveness of the nickel, hence, can balance the structural perturbations caused by the active Mn^{2+}/Mn^{3+} during the redox processes.[197] Even though all these hexacyanoferrate compounds demonstrate potential electrochemical properties and some of the compounds are commercialized; the production of the compound and/or their precursors could trigger environmental issues, which needs to be fixed. Besides, another major concern is the problems associated with the controlling of the water content in the sample, which is an unfavorable attribute for the storage property of these samples.

4.4.3 Sulfates

One of the best commercially available battery materials for LIBs is LiFePO$_4$. However, the sodium analogue of this material suffers from low operating potential and hence poor specific energy density. Substituting (PO$_4$)$^{3-}$ polyanion with a more electronegative (SO$_4$)$^{2-}$ group is an effective strategy to enhance the operating potential and thus to resolve the poor energy density issue associated with the existing SIBs.[198] Recently, a study reported that the Na$_2$Fe$_2$(SO$_4$)$_3$ cathode material could achieve a high Fe redox potential (~3.8 V) vs Na/Na$^+$.[199] This study also demonstrated that the material could exhibit good electrochemical stability and sodium ion transportation. Further, this alluaudite-type Na$_2$Fe$_2$(SO$_4$)$_3$, could deliver a specific discharge capacity of ~100 mAh g^{-1} with good cycle and rate performances. The theoretical energy density of this material is estimated to be more than 540 Wh kg^{-1} (vs Na/Na$^+$) because of the higher operating potential, which ensures that SIBs could compete with the state-of-the-art Li-ion batteries in a matter of time.[199] The kröhnkite-type Na$_2$Fe(SO$_4$)$_2$·2H$_2$O could deliver a reversible capacity of 70 mAh g^{-1} at a C-rate of 0.05C with an operating potential of ~3.25 V and a cycling stability of 80% after 20 cycles.[200] Another study on the same material reported a specific capacity of 72 mAh g^{-1} at 0.05C at an operating voltage of ~3.234 V and a better cycling stability.[201] A rhombohedral NASICON-structured iron(III) sulfate material prepared via ball milling technique has demonstrated a capacity of 65 mAh g^{-1} with a capacity retention of 80% for over 400 cycles with an operating potential of 3.2 V for a Fe^{2+}/Fe^{3+} redox couple.[202] The eldfellite NaFe(SO$_4$)$_2$ prepared via solution route showed a specific discharge capacity of ~85 mA h g^{-1} at 0.05C with an operating potential of 3 V.[203] Another study reported that manganese substitution in Na$_{2.5}$(Fe$_{1-y}$Mn$_y$)$_{1.75}$(SO$_4$)$_3$ (y = 0, 0.25, 0.5, 0.75, and 1.0) could increase the potential of a Fe^{3+}/Fe^{2+} redox couple, but with a simple capacity compromise, due to the presence of inactive Mn^{2+} units.[203] Recently, fluorinated polyanion compounds with tavorite-type structure have also received much attention.[205-207] Where the incorporation of fluoride in the sulfate compounds brings out a charge difference and alteration of the dimensionality of the lattice along with a turned redox potential. This crystal structure has linear chains of corner-sharing octahedral FeO$_4$F$_2$ which spread along the c-axis and are connected by corner-sharing tetrahedral SO$_4$ along the a and b axes to create a cavernous network, with two large intersecting tunnels for fast ion transportation, thus it can hold and allow multi-dimensional transport pathways for larger ions like sodium (Figure 4.4d).[208,209] This makes sure that enhanced electrochemical properties can even be achieved with submicron particles; this is a favorable attribute to the material processing. Only enhancement in ionic conductivity could not lead to the superior overall electrochemical performances. For example, the fluoro-sulphate polyanion material in its pure state has low rate capability due to its poor electronic conductivity. This problem will become worse in the case of a sulfate-based system, due to its lower decomposition temperature, hence the *in-situ* coating of conducting carbons and/or cationic doping (to improve the electronic conductivity) could be difficult. This could be one of the reasons why this particular category of materials has not been widely investigated for battery electrodes. This problem, however, can be addressed by employing a combination of ball mill and solid-state technique to synthesize a sodium-based fluoro-sulphate polyanion (NaFeSO$_4$F) with optimal electrochemical properties. Where the CNTs were added along with the NaFeSO$_4$F precursors to ensure the intact and effective contact of the final product (NaFeSO$_4$F) with the CNTs. The CNTs lined NaFeSO$_4$F material demonstrated improved electronic conductivity, high specific discharge capacity (~110 mAh g^{-1} at 0.1C), and good cycling stability of 200 cycles with more than 91% capacity retention for the Fe^{2+}/Fe^{3+} redox couple.[210]

4.5 Organic Materials

Compared with inorganic materials for SIBs, the organic compounds could demonstrate some unique properties, such as structural diversity, less expensive, good safety, recyclable nature, and flexibility. In general, there are two kinds of organic materials for SIBs. The first category is the metal organic materials and the second category is the metal-free organic materials. The metal hexacyanometalates with the general chemical formula of A$_x$MM^1(CN)$_6$ (A = Na, K; M and M^1 = Fe, Co, Mn, Ni, In) come under metal organic materials. This material shows cubic structure, where the metal ions are located at the corners and with cyanide units linked along the cube edges. Also, these metal ions are coordinated octahedrally by the nitrogen or carbon end of the cyanide units. These crystal structures permit reversible intercalation/de-intercalation mechanisms of alkali-metal ions, such as Li, Na, K, Ca, etc., due to their suitable interstitial sites. Among different hexacyanometalates, the hexacyanoferrates are the most explored compounds for SIBs (the detailed discussion can be found in Section 4.3.2: Hexacyanoferrates). Thus, in this section the main focus will be on metal-free organic compounds.

4.5.1 Non-polymeric Materials

Organic compounds obtained from biomass are a suitable choice for the future SIBs application, due to its environmentally friendly and low-cost nature.[211,212] This compound's solubility in organic electrolytes, poor thermal stability, sluggish kinetics, and low electronic conductivity limits its applications in LIBs and SIBs.[213] In recent years, different organic materials, including aromatic carbonyl derivatives and pethidine derivatives, however, have been explored as cathode systems for SIBs. Some of these organic cathodes have exhibited potential electrochemical properties, such as high specific discharge capacities, high average operating voltages, etc. For instance, the disodium rhodizonate material with a chemical formula of $Na_2C_6O_6$ has demonstrated an excellent high discharge capacity of ~270 mA h g^{-1} in 1 m $NaClO_4$/PC electrolyte in the potential window of 1.0–2.9 V. This high capacity value could be attributed to the intercalation of more than two sodium ions per $Na_2C_6O_6$ molecule. When this material assembled as a full-cell against pre-doped hard carbon, a second specific discharge capacity of 179 mAh g^{-1} with a cycling stability of 85% after the 40th cycle was observed.[214] These carbonyl-based organic materials have shown a layered structure, thus the structure is considered to be appropriate for sodium ion insertion/extraction processes. The various reaction paths in the sodiation/desodiation of the $Na_2C_6O_6$ compound are shown in Figure 4.5a.[215] Another study conducted on the $Na_2C_6O_6$ compounds revealed obvious size-dependent sodium-ion-storage properties. Among different structured (microbulk, microrod, and nanorod) $Na_2C_6O_6$ compounds, the nanorods delivered the best electrochemical performances, that is a specific capacity of ~190 mA h g^{-1} at 0.1C with a cycling stability of >90% over 100 cycles. Also, a promising rate capability of 10C with 50% of the capacity even at 80°C was reported for this compound due to its good reaction kinetics. Thus, such a biomass-derived organic cathode material prepared using an anti-solvent technique demonstrates a potential route toward high-performing organic-based cathode systems for beyond LIBs, due to its sustainability, low-cost, high electrochemical properties, and thermal stability.[215] The solubility of these biomass-derived cathodes in an organic electrolyte can be controlled by selective molecular design and through proper selection of the potential window. Using these ways, the dissolution of material into the electrolyte can be minimized, leading to the improved cycling performances. For instance, a study reported that the Na_6C_6O designed using keto-carbonyl groups could demonstrate specific discharge capacities of 173.5 and 115 mA h g^{-1} at current densities of 50 mA g^{-1} and at 5000 mA g^{-1}, respectively, in the voltage window of 1.0–3.0 V. This designed material also presented a long cycle life of over 1500 cycles with ~90% capacity retention at a current density of 1000 mA g^{-1}.[216]

110   Advanced Materials for Sodium Ion Storage

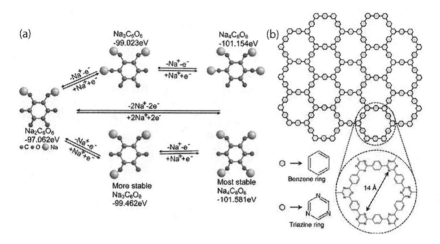

FIGURE 4.5
(a) Various reaction paths in the sodiation/desodiation of the Na$_2$C$_6$O$_6$ materials. [Reprinted with permission from *Nano Lett.* 2016, 16, 3329. Copyright (2016) American Chemical Society.] and (b) schematic representation of BPOE having a porous-honeycomb structure. The 2D structure of the bipolar porous organic electrode (BPOE) is comprised of triazine and benzene rings. [Reprinted with permission from Springer Nature: *Nat. Commun.*, 2013, 4, 1485. Copyright (2013).]

Another study demonstrated that the extension of the π-conjugated system was an effective way to enhance the electrochemical performances of organic cathodes.[216] It is reported that the π-conjugated materials can facilitate the sodium ion transport, stabilize the electrochemical states, and enhance the sodium ion insertion/extraction processes. Recently, such a π-conjugated sodium 4, 4′-stilbene-dicarboxylate, cathode material showed a discharge specific capacity of 220 mA h g^{-1} at 50 mA g^{-1} in the potential window of 0.1–2.5 V, with very high rate capabilities of 105 and 72 mA h g^{-1} at current densities of 2 A g^{-1} and 10 A g^{-1}, respectively. This material also demonstrated a reasonably good cycling stability of >70% after 400 cycles of charging/discharging.[217] Another potential organic cathode material is 3,4,9,10-perylene-tetracarboxylicacid-dianhydride (PTCDA) (C$_{24}$H$_8$O$_6$); also called as "red 224." This is an organic pigment with an aromatic core and two anhydride groups. A recent study reported that this material could deliver a discharge specific capacity of ~140 mA h g^{-1} at a current density of 10 mA g^{-1} with a good cycling stability over 200 cycles in the potential window of 1.0–3.0 V and a high rate capability of 91 mA h g^{-1} at a current density of 1000 mA g^{-1}.[218] Another commonly investigated organic compound for SIBs is tetra-sodium salt of 2,5-dihydroxyterephthalic acid (Na$_4$DHTPA or Na$_4$C$_8$H$_2$O$_6$). This material can be utilized both as an anode and cathode by selecting the appropriate potential window that is Na$_4$DHTPA at 1.6–2.8 V acts as cathode and at 0.1–1.8 V performs as anode (vs Na$^+$/Na). Both the anode and cathode could deliver a discharge capacity of ~180 mA h g^{-1} with a two-electron process.[219] The full-cell fabricated out of Na$_4$DHTPA as anode and as cathode material showed an average potential of 1.8 V and an energy density of ~65 W h kg^{-1}. However, the aforementioned quinone-based cathodes always suffered from low operating potential (<2.5 V; vs Na/Na$^+$), which restricts their further practical application as high energy-density cathodes for SIBs. A promising approach to mitigate this issue is to introduce electronegative elements like Cl or Br, to lower the lowest unoccupied molecular orbital (LUMO) levels and in turn enhance the sodium storage properties of the quinone-derivatives. For instance, a C$_6$Cl$_4$O$_2$/mesoporous carbon composite system could demonstrate a high specific discharge capacity of 161 mA h g^{-1} at an average operating voltage of ~2.72 V with a superior specific energy density of 420 W h kg^{-1}. This is one of the highest energy densities reported yet for organic materials as a cathode in SIBs.[220] Even though the organic compounds have been widely explored, the electrode material dissolution into the organic electrolyte, leading to poor capacity retention, is still a bottleneck for the practical applications. Thus, yet to be discovered, an organic material which is insoluble in an organic electrolyte to develop a potential low-cost, high-energy, and long-lasting cathode in SIBs.

4.5.2 Polymeric Materials

The electro-active p-type polymers are considered as a promising cathode material for SIBs applications, due to their superior properties such as high redox activity (lead to high capacity), flexible matrixes (tolerate the volume changes during redox processes), and improved insolubility in an organic electrolyte (lead to the high capacity retention).[221] The charge/discharge process of these polymer systems is due to the insertion/release of the anions like PF6$^-$ and ClO$_4^-$ in the electrolyte into and from the flexible host polymer networks. This mechanism is commonly known as the anion-insertion process.[222] A study on amorphous oligopyrene (OPr, 3–4 pyrene units) for SIBs demonstrated a voltage plateau of 3.5 V in a NaClO$_4$ electrolyte, with an initial specific discharge capacity of 121 mA h g^{-1} at the current density of 20 mA g^{-1} and high 1st discharge energy density of 423 W h kg^{-1}.[223] One of the potential ways to get good electrochemical performances for organic cathodes is to modify the redox active small organic molecules into the form of a polymer framework. For instance, a recent study developed excellent organic materials using modified aromatic carbonyl-derivative polyimides (PIs), such as pyromellitic dianhydride (PMDA), 1,4,5,8-naphthalenetetracarboxylic dianhydride, and perylene 3,4,9,10-tetracarboxylic dianhydride (PTCDA).[224]

Among them, PTCDA-PIs cathode demonstrated the best electrochemical properties, such as high power density of 20.99 kW kg^{-1}, energy density of 285 W h kg^{-1}, and good cycle life with 87.5% capacity retention after 5000 cycles, due to its low solubility in an electrolyte and the lowest LUMO energy level.[224] The conventional polymerization process, however, introduces a large portion of redox-inactive ligands for the intra-molecular connections, resulting in a reduction in active mass and thus the specific capacity. The preparation of a π-conjugated polymer system which is self-connected by small redox-active units can be beneficial to address the above issue. For instance, the π-conjugated poly(anthraquinonyl imide)s (PAQIs) prepared from pyromellitic dianhydride (or 1,4,5,8-naphthalenetetracarboxylic dianhydride) and 1,4-diaminoanthraquinone (or 1,5-diaminoanthraquinone) demonstrated a high specific discharge capacity of 190 mAh g^{-1} with a cycle life of ~93% capacity retention after 150 cycles.[225] A new polymer cathode material, poly(benzoquinonyl sulfide) (PBQS), is shown to have an excellent energy density of 557 W h kg^{-1} at a current density of 268 mA h g^{-1} with an operating potential of 2.08 V.[226] Another superior cathode material is a porous-honeycomb-structured bipolar polymeric framework, which consists of benzene and triazine rings in a 2D structure (Figure 4.5b). This cathode material could deliver excellent electrochemical performances such as high power density (10 kW kg^{-1}), superior energy density (500 Wh kg^{-1}), and excellent cycling stability (7,000 cycles with ~80% capacity retention).[226] The electrochemical performances of this cost-effective material exceed those of most inorganic lithium and sodium cathodes.

4.6 Conclusions and Future Outlook

Extensive studies on cathode materials for SIBs have been carried out by many scientific groups. Some of these studies demonstrated excellent electrochemical performance in terms of specific capacity, cycle life, rate capability, and specific energy. However, most of the full-cell performances did not meet the required properties for the commercialization. The works showed good full-cell performances, however, required pre-sodiation process on the anode/cathode side, which brought difficulty for the large-scale preparation. Thus, whether SIBs electrochemical performances will eventuality meet with that of LIBs is still an open debate; however, definitely, successful investigation of suitable cathode materials could make SIBs more competitive for future commercial applications. In this chapter, a variety of the electrochemical properties of promising cathode materials have been discussed, which include layered transition-metal oxides, NASICON-based materials, phosphates, sulfates, pyrophosphates, hexacyanometalates, organic materials, etc. Transition-metal oxides: The highlight of this material is its high specific capacity (it could deliver over 200 mA h g^{-1}), and thus high energy density (could achieve over 600 W h kg^{-1}). Apart from its low operating voltage (~2.7–3.0 V), this cathode system, in general, suffers from poor cycling performance and rate capability, because of its lattice expansion and multi-phase transitions. At present, by comparing the electrochemical performances and the feasibility of the material synthesis, transition-metal oxides are the potential cathode candidate for the current SIBs. However, the stability of these materials in air has always been a concern for layered transition-metal oxide cathodes. The O3-type cathode could deliver high specific capacity, but it suffers from poor air stability and low cycling performances. The P2-type materials could demonstrate good cycling and air stability because the sodium ion occupied at a larger trigonal prismatic site, which is beneficial for the sodium ion transport. Further, the stability in air could be improved by an effective reductive carbon encapsulation. Phosphates: The simple phosphates in general can demonstrate a good cycling life, but their main drawbacks are low specific discharge capacity (<110 mA h g^{-1}), and hence the poor specific energy (<350 W h kg^{-1}), due to the high molecular weight and sluggish sodium ion insertion/extraction process. The pyrophosphates can attain high operating voltages, but they show poor specific capacities in the range of ~90 mA h g^{-1}, with a specific energy density of about 350 W h kg^{-1}. Sometimes these pyrophosphates could demonstrate very high potential of over 4.3 V, which is beyond the stable potential window of the commercially available electrolyte, leading to capacity fading. While the fluorophosphate cathodes could deliver potential electrochemical properties such as high specific discharge capacities of over 120 mA h g^{-1}, high operating potential of about 3.9 V, promising energy densities in the range of 400–500 W h kg^{-1}, and good cycling performances, due to their superior inductive effects. Hexacyanometalates and organic materials: The hexacyanometalates and organic compounds can demonstrate a wide range of electrochemical properties by modulating the molecular structure. For the case of organic materials, the electrochemical performances depend strongly on the carbon additive. Hexacyanometalates compounds could demonstrate high capacity and energy densities in the range of 500–600 W h kg^{-1} with limited cycle life. Thus, for most of the investigated cathodes, superior electrochemical properties balanced by some inferior electrochemical performance. To avoid the safety and dissolution issues, replacing the organic liquid electrolyte with solid-state electrolyte with good ionic conductivity would be an ideal solution for organic materials. Also, the aqueous electrolyte SIBs are considered to be a potential future option, because of their enhanced safety, low-cost nature, ease of production, and the high ionic conductivity of the water-based electrolyte. For most of the cathodes, the electrochemical performances could be enhanced by selecting the appropriate approach, such as doping, conducting carbon coating, crystal-structure design, nanostructuring, compositing, and hybridizations. The reaction kinetics of the cathode material is generally more prone to the crystal structure and elemental composition. Hence, the electrochemical properties of these materials can be enhanced by tuning the abovementioned characteristics. For monoanionic material, the partial substitution of cation could enhance the electrochemical properties such as specific capacity, cycling stability, and conductivity. Also, the substitution with some redox inactive metals, like Mg, Ca, and Sb, has shown to enhance the structural stability, thus the cycling performance of the material. The Mn doping will be an effective way to suppress the Jahn-Teller-induced structural changes and ordering of sodium ion; resulting in improved sodium ion transport and cycling stability. The Li partial substitution in transition oxide material can suppress the phase transformation during the charge/discharge process, leading to superior cycling performances. For most of the polyanionic material, the anion-substitution (like fluorine) is an effective way to enhance the operating potential. Oxygen substitution is also observed to be an efficient route to introduce multi-electron transfer to improve the electrochemical properties. Also, different mixtures of polyanionic units could bring out new open crystal structures with adjustable redox couples and complex multi-dimensional pathways for rapid sodium ion transport kinetics. For the organic compounds, their sluggish charge transport kinetics and dissolution of electrode material in electrolyte restrict their practical SIB applications. Thus, wrapping the organic material with appropriate carbon framework is a very efficient technique to enhance the charge-transfer kinetics. Another potential way to enhance the electrochemical properties of cathodes in SIBs is nanostructuring. Nowadays, the cathode materials can be easily nanostructured in any morphology and dimensions, leading to improved electrochemical performances. Another way to enhance the electrochemical performances of the material is to understand the fundamental problems that are responsible for the inferior properties. This can be achieved through many *in situ* and *ex situ* advanced characterization techniques, such as Transmission Electron Microscopy (TEM), Nuclear Magnetic Resonance (NMR), X-ray powder diffraction (XRD), Near Edge X-ray Absorption Fine Structure (NEXAFS), Extended X-ray Absorption Fine Structure (EXAFS), etc. *In-situ* studies like neutron and synchrotron diffraction techniques are some of the effective tools to understand the structural deformation of the anode/cathode systems during charge/discharge processes. These techniques are also useful to thoroughly investigate the root cause for the irreversible capacity loss in the electrode and thus to find out the necessary measures to overcome this drawback. Thus, these *in situ* characterizations are very critical for the development of potential electrodes for future SIBs. A detailed computational simulation can also be useful to design new cathode materials and enhance the electrochemical performances of the existing materials. To fabricate a potential SIB for practical applications, researches should focus to develop new compositions and strategies for the preparation of material. Also, the fundamental research should be focused to bring the SIBs from laboratory bench to open market. Thus, in fact, both the fundamental and industrial research should meet to bring out the SIBs in real applications. Until now, the full-cell performances of SIBs are not yet close to that of commercially available LIBs. However, companies like United Kingdom-based Faradion and USA-based Aquion Energy are extensively working on SIBs [22]. For instance, they have presented e-bikes based on SIBs as a proof of concept. This has opened much room to explore and optimize a new class of electrode materials for the successful commercialization of SIBs in the near future.

References

1. Zhang, X., Zhang, Z., Yao, S., Chen, A., Zhao, X., Zhou, Z. An effective method to screen sodium-based layered materials for sodium ion batteries. *NPJ Computational Materials* **2018**, *4* (1).
2. Dai, Z., Mani, U., Tan, H. T., Yan, Q. Advanced cathode materials for sodium-ion batteries: What determines our choices? *Small Methods* **2017**, *1* (5), 1700098.
3. Peters, J., Buchholz, D., Passerini, S., Weil, M. Life cycle assessment of sodium-ion batteries. *Energy Environ. Sci.* **2016**, *9* (5), 1744.
4. Islam, M. S., Fisher, C. A. J. Lithium and sodium battery cathode materials: Computational insights into voltage, diffusion and nanostructural properties. *Chem. Soc. Rev.* **2014**, *43* (1), 185.
5. Zelang, J., Wenze, H., Xia, L., Huaixin, Y., Yong-Sheng, H., Jing, Z., Zhibin, Z., Jianqi, L., Wen, C., Dongfeng, C. et al. Superior electrochemical performance and storage mechanism of $Na_3V_2(PO_4)_3$ cathode for room-temperature sodium-ion batteries. *Adv. Energy Mater.* **2013**, *3* (2), 156.
6. He, H., Jin, G., Wang, H., Huang, X., Chen, Z., Sun, D., Tang, Y. Annealed NaV_3O_8 nanowires with good cycling stability as a novel cathode for Na-ion batteries. *J. Mater. Chem. A* **2014**, *2* (10), 3563.
7. Shuang, Y., Yong-Bing, L., Dan, X., De-Long, M., Sai, W., Xiao-Hong, Y., Zhan-Yi, C., Xin-Bo, Z. Pure single-crystalline $Na_{1.1}V_3O_{7.9}$ nanobelts as superior cathode materials for rechargeable sodium-ion batteries. *Adv. Sci.* **2015**, *2* (3), 1400018.
8. Whittingham, M. S. Chemistry of intercalation compounds: Metal guests in chalcogenide hosts. *Prog. Solid State Ch.* **1978**, *12* (1), 59.
9. Braconnier, J.-J., Delmas, C., Fouassier, C., Hagenmuller, P. Comportement electrochimique des phases Na_xCoO_2. *Mater. Res. Bull.* **1980**, *15* (12), 7.
10. Yabuuchi, N., Kubota, K., Dahbi, M., Komaba, S. Research development on sodium-ion batteries. *Chem. Rev.* **2014**, *114* (23), 11636.
11. Doeff, M. M., Peng, M. Y., Ma, Y., Jonghe, L. C. D. Orthorhombic Na_xMnO_2 as a cathode material for secondary sodium and lithium polymer batteries. *J. Electrochem. Soc.* **1994**, *141* (11), 3.
12. Pan, H., Hu, Y.-S., Chen, L. Room-temperature stationary sodium-ion batteries for large-scale electric energy storage. *Energy Environ. Sci.* **2013**, *6* (8), 2338.
13. Yunming, L., Zhenzhong, Y., Shuyin, X., Linqin, M., Lin, G., Yong-Sheng, H., Hong, L., Liquan, C. Air-stable copper-based P2-$Na_{7/9}Cu_{2/9}Fe_{1/9}Mn_{2/3}O_2$ as a new positive electrode material for sodium-ion batteries. *Adv. Sci.* **2015**, *2* (6), 1500031.

14. Ding, J. J., Zhou, Y. N., Sun, Q., Yu, X. Q., Yang, X. Q., Fu, Z. W. Electrochemical properties of P2-phase Na$_{0.74}$CoO$_2$ compounds as cathode material for rechargeable sodium-ion batteries. *Electrochim. Acta* **2013**, *87*, 388.
15. Oh, S.-M., Myung, S.-T., Yoon, C. S., Lu, J., Hassoun, J., Scrosati, B., Amine, K., Sun, Y.-K. Advanced Na[Ni$_{0.25}$Fe$_{0.5}$Mn$_{0.25}$]O$_2$/C–Fe$_3$O$_4$ sodium-ion batteries using EMS electrolyte for energy storage. *Nano Lett.* **2014**, *14* (3), 1620.
16. Linqin, M., Shuyin, X., Yunming, L., Yong-Sheng, H., Hong, L., Liquan, C., Xuejie, H. Prototype sodium-ion batteries using an air-stable and Co/Ni-Free O$_3$-layered metal oxide cathode. *Adv. Mater.* **2015**, *27* (43), 6928.
17. Han, M. H., Gonzalo, E., Singh, G., Rojo, T. A comprehensive review of sodium layered oxides: Powerful cathodes for Na-ion batteries. *Energy Environ. Sci.* **2015**, *8* (1), 81.
18. Shaohua, G., Haijun, Y., Zelang, J., Pan, L., Yanbei, Z., Xianwei, G., Mingwei, C., Masayoshi, I., Haoshen, Z. A high-capacity, low-cost layered sodium manganese oxide material as cathode for sodium-ion batteries. *ChemSusChem* **2014**, *7* (8), 2115.
19. Billaud, J., Clément, R. J., Armstrong, A. R., Canales-Vázquez, J., Rozier, P., Grey, C. P., Bruce, P. G. β-NaMnO$_2$: A high-performance cathode for sodium-ion batteries. *J. Am. Chem. Soc.* **2014**, *136* (49), 17243.
20. Yabuuchi, N., Yoshida, H., Komaba, S. Crystal structures and electrode performance of alpha-NaFeO$_2$ for rechargeable sodium batteries. *Electrochemistry* **2012**, *80* (10), 716.
21. Zhao, J., Zhao, L., Dimov, N., Okada, S., Nishida, T. Electrochemical and thermal properties of α-NaFeO$_2$ cathode for Na-ion batteries. *J. Electrochem. Soc.* **2013**, *160* (5), A3077.
22. Yabuuchi, N., Kajiyama, M., Iwatate, J., Nishikawa, H., Hitomi, S., Okuyama, R., Usui, R., Yamada, Y., Komaba, S. P2-type Na$_x$[Fe$_{1/2}$Mn$_{1/2}$]O$_2$ made from earth-abundant elements for rechargeable Na batteries. *Nat. Mater.* **2012**, *11*, 512.
23. Thorne, J. S., Dunlap, R. A., Obrovac, M. N. Structure and electrochemistry of Na$_x$Fe$_x$Mn$_{1-x}$O$_2$ (1.0 ≤ x ≤ 0.5) for Na-ion battery positive electrodes. *J. Electrochem. Soc.* **2013**, *160* (2), A361.
24. Lee, E., Brown, D. E., Alp, E. E., Ren, Y., Lu, J., Woo, J.-J., Johnson, C. S. New insights into the performance degradation of Fe-based layered oxides in sodium-ion batteries: Instability of Fe^{3+}/Fe^{4+} redox in α-NaFeO$_2$. *Chem. Mater.* **2015**, *27* (19), 6755.
25. Zhang, Z.-J., Wang, J.-Z., Chou, S.-L., Liu, H.-K., Ozawa, K., Li, H.-J. Polypyrrole-coated α-LiFeO$_2$ nanocomposite with enhanced electrochemical properties for lithium-ion batteries. *Electrochim. Acta* **2013**, *108*, 820.
26. Komaba, S., Takei, C., Nakayama, T., Ogata, A., Yabuuchi, N. Electrochemical intercalation activity of layered NaCrO$_2$ vs. LiCrO$_2$. *Electrochem. Commun.* **2010**, *12* (3), 355.
27. Ding, J.-J., Zhou, Y.-N., Sun, Q., Fu, Z.-W. Cycle performance improvement of NaCrO$_2$ cathode by carbon coating for sodium ion batteries. *Electrochem. Commun.* **2012**, *22*, 85.
28. Chen, C.-Y., Matsumoto, K., Nohira, T., Hagiwara, R., Fukunaga, A., Sakai, S., Nitta, K., Inazawa, S. Electrochemical and structural investigation of NaCrO$_2$ as a positive electrode for sodium secondary battery using inorganic ionic liquid NaFSA–KFSA. *J. Power Sources* **2013**, *237*, 52.
29. Miyazaki, S., Kikkawa, S., Koizumi, M. Chemical and electrochemical deintercalations of the layered compounds LiMO$_2$ (M = Cr, Co) and NaM'O$_2$ (M' Cr, Fe, Co, Ni). *Synthetic Metals* **1983**, *6*, 211.

30. Xia, X., Dahn, J. R. NaCrO$_2$ is a fundamentally safe positive electrode material for sodium-ion batteries with liquid electrolytes. *Electrochem. Solid State Lett.* **2011**, *15* (1), A1.
31. Yu, C.-Y., Park, J.-S., Jung, H.-G., Chung, K.-Y., Aurbach, D., Sun, Y.-K., Myung, S.-T. NaCrO$_2$ cathode for high-rate sodium-ion batteries. *Energy Environ. Sci.* **2015**, *8* (7), 2019.
32. Bo, S.-H., Li, X., Toumar, A. J., Ceder, G. Layered-to-rock-salt transformation in desodiated Na$_x$CrO$_2$ (x 0.4). *Chem. Mater.* **2016**, *28* (5), 1419.
33. Molenda, J., StokŁlosa, A. Electronic and electrochemical properties of nickel bronze, NaxNiO$_2$. *Solid State Ion.* **1990**, *38* (1), 1.
34. Vassilaras, P., Ma, X., Li, X., Ceder, G. Electrochemical properties of monoclinic NaNiO$_2$. *J. Electrochem. Soc.* **2013**, *160* (2), A207.
35. Komaba, S., Yabuuchi, N., Nakayama, T., Ogata, A., Ishikawa, T., Nakai, I. Study on the reversible electrode reaction of Na$_{1-x}$Ni$_{0.5}$Mn$_{0.5}$O$_2$ for a rechargeable sodium-ion battery. *Inorg. Chem.* **2012**, *51* (11), 6211.
36. Wang, X., Gao, Y., Shen, X., Li, Y., Kong, Q., Lee, S., Wang, Z., Yu, R., Hu, Y.-S., Chen, L. Anti-P2 structured Na$_{0.5}$NbO$_2$ and its negative strain effect. *Energy Environ. Sci.* **2015**, *8* (9), 2753.
37. Tamaru, M., Wang, X., Okubo, M., Yamada, A. Layered Na$_2$RuO$_3$ as a cathode material for Na-ion batteries. *Electrochem. Commun.* **2013**, *33*, 23.
38. Hasa, I., Buchholz, D., Passerini, S., Hassoun, J. A comparative study of layered transition metal oxide cathodes for application in sodium-ion battery. *ACS Appl. Mater. Interfaces* **2015**, *7* (9), 5206.
39. Uchaker, E., Jin, H., Yi, P., Cao, G. Elucidating the role of defects for electrochemical intercalation in sodium vanadium oxide. *Chem. Mater.* **2015**, *27* (20), 7082.
40. Yuan, D. D., Wang, Y. X., Cao, Y. L., Ai, X. P., Yang, H. X. Improved electrochemical performance of Fe-substituted NaNi$_{0.5}$Mn$_{0.5}$O$_2$ cathode materials for sodium-ion batteries. *ACS Appl. Mater. Interfaces* **2015**, *7* (16), 8585.
41. Wang, H., Yang, B., Liao, X.-Z., Xu, J., Yang, D., He, Y.-S., Ma, Z.-F. Electrochemical properties of P2-Na$_{2/3}$[Ni$_{1/3}$Mn$_{2/3}$]O$_2$ cathode material for sodium ion batteries when cycled in different voltage ranges. *Electrochim. Acta* **2013**, *113*, 200.
42. Zhu, Y.-E., Qi, X., Chen, X., Zhou, X., Zhang, X., Wei, J., Hu, Y., Zhou, Z. A P2-Na$_{0.67}$Co$_{0.5}$Mn$_{0.5}$O$_2$ cathode material with excellent rate capability and cycling stability for sodium ion batteries. *J. Mater. Chem. A* **2016**, *4* (28), 11103.
43. Kalluri, S., Seng, K. H., Pang, W. K., Guo, Z., Chen, Z., Liu, H.-K., Dou, S. X. Electrospun P2-type Na$_{2/3}$(Fe$_{1/2}$Mn$_{1/2}$)O$_2$ hierarchical nanofibers as cathode material for sodium-ion batteries. *ACS Appl. Mater. Interfaces* **2014**, *6* (12), 8953.
44. Jung, Y. H., Christiansen, A. S., Johnsen, R. E., Norby, P., Kim, D. K. In situ X-ray diffraction studies on structural changes of a P2 layered material during electrochemical desodiation/sodiation. *Adv. Funct. Mater.* **2015**, *25* (21), 3227.
45. Sharma, N., Gonzalo, E., Pramudita, J. C., Han, M. H., Brand, H. E. A., Hart, J. N., Pang, W. K., Guo, Z., Rojo, T. The unique structural evolution of the O$_3$-phase Na$_{2/3}$Fe$_{2/3}$Mn$_{1/3}$O$_2$ during high rate charge/discharge: A sodium-centred perspective. *Adv. Funct. Mater.* **2015**, *25* (31), 4994.
46. Bucher, N., Hartung, S., Franklin, J. B., Wise, A. M., Lim, L. Y., Chen, H.-Y., Weker, J. N., Toney, M. F., Srinivasan, M. P2–Na$_x$Co$_y$Mn$_{1-y}$O$_2$ (y = 0, 0.1) as cathode materials in sodium-ion batteries: Effects of doping and morphology to enhance cycling stability. *Chem. Mater.* **2016**, *28* (7), 2041.

47. Ivana, H., Daniel, B., Stefano, P., Bruno, S., Jusef, H. High performance $Na_{0.5}[Ni_{0.23}Fe_{0.13}Mn_{0.63}]O_2$ cathode for sodium-ion batteries. *Adv. Energy Mater.* **2014**, *4* (15), 1400083.
48. Hwang, J.-Y., Oh, S.-M., Myung, S.-T., Chung, K. Y., Belharouak, I., Sun, Y.-K. Radially aligned hierarchical columnar structure as a cathode material for high energy density sodium-ion batteries. *Nat. Commun.* **2015**, *6*, 6865.
49. Sun, X., Jin, Y., Zhang, C.-Y., Wen, J.-W., Shao, Y., Zang, Y., Chen, C.-H. $Na[Ni_{0.4}Fe_{0.2}Mn_{0.4-x}Ti_x]O_2$: A cathode of high capacity and superior cyclability for Na-ion batteries. *J. Mater. Chem. A* **2014**, *2* (41), 17268.
50. Singh, G., Aguesse, F., Otaegui, L., Goikolea, E., Gonzalo, E., Segalini, J., Rojo, T. Electrochemical performance of $NaFe_x(Ni_{0.5}Ti_{0.5})_{1-x}O_2$ ($x = 0.2$ and $x = 0.4$) cathode for sodium-ion battery. *J. Power Sources* **2015**, *273*, 333.
51. Yu, H., Guo, S., Zhu, Y., Ishida, M., Zhou, H. Novel titanium-based O3-type $NaTi_{0.5}Ni_{0.5}O_2$ as a cathode material for sodium ion batteries. *Chem. Commun.* **2014**, *50* (4), 457.
52. Han, M. H., Gonzalo, E., Sharma, N., López del Amo, J. M., Armand, M., Avdeev, M., Saiz Garitaonandia, J. J., Rojo, T. High-performance P2-phase $Na_{2/3}Mn_{0.8}Fe_{0.1}Ti_{0.1}O_2$ cathode material for ambient-temperature sodium-ion batteries. *Chem. Mater.* **2016**, *28* (1), 106.
53. Wang, Y., Xiao, R., Hu, Y.-S., Avdeev, M., Chen, L. P2-$Na_{0.6}[Cr_{0.6}Ti_{0.4}]O_2$ cation-disordered electrode for high-rate symmetric rechargeable sodium-ion batteries. *Nat. Commun.* **2015**, *6*, 6954.
54. Shaohua, G., Pan, L., Yang, S., Kai, Z., Jin, Y., Mingwei, C., Masayoshi, I., Haoshen, Z. A high-voltage and ultralong-life sodium full cell for stationary energy storage. *Angewandte Chemie International Edition* **2015**, *54* (40), 11701.
55. Han, J., Niu, Y., Zhang, Y., Jiang, J., Bao, S.-j., Xu, M. Evaluation of O3-type $Na_{0.8}Ni_{0.6}Sb_{0.4}O_2$ as cathode materials for sodium-ion batteries. *J. Solid State Electrochem.* **2016**, *20* (8), 2331.
56. Billaud, J., Singh, G., Armstrong, A. R., Gonzalo, E., Roddatis, V., Armand, M., Rojo, T., Bruce, P. G. $Na_{0.67}Mn_{1-x}Mg_xO_2$ ($0 \leq x \leq 0.2$): A high capacity cathode for sodium-ion batteries. *Energy Environ. Sci.* **2014**, *7* (4), 1387.
57. Ma, J., Bo, S.-H., Wu, L., Zhu, Y., Grey, C. P., Khalifah, P. G. Ordered and disordered polymorphs of $Na(Ni_{2/3}Sb_{1/3})O_2$: Honeycomb-ordered cathodes for Na-ion batteries. *Chem. Mater.* **2015**, *27* (7), 2387.
58. Xu, J., Lee, D. H., Clément, R. J., Yu, X., Leskes, M., Pell, A. J., Pintacuda, G., Yang, X.-Q., Grey, C. P., Meng, Y. S. Identifying the critical role of Li substitution in P2-$Na_x[Li_yNi_zMn_{1-y-z}]O_2$ ($0 < x, y, z < 1$) intercalation cathode materials for high-energy Na-ion batteries. *Chem. Mater.* **2014**, *26* (2), 1260.
59. Oh, S.-M., Myung, S.-T., Hwang, J.-Y., Scrosati, B., Amine, K., Sun, Y.-K. High capacity O3-type $Na[Li_{0.05}(Ni_{0.25}Fe_{0.25}Mn_{0.5})_{0.95}]O_2$ cathode for sodium ion batteries. *Chem. Mater.* **2014**, *26* (21), 6165.
60. Shaohua, G., Pan, L., Haijun, Y., Yanbei, Z., Mingwei, C., Masayoshi, I., Haoshen, Z. A layered P2- and O3-type composite as a high-energy cathode for rechargeable sodium-ion batteries. *Angewandte Chemie International Edition* **2015**, *54* (20), 5894.
61. Su, D., Ahn, H.-J., Wang, G. Hydrothermal synthesis of α-MnO_2 and β-MnO_2 nanorods as high capacity cathode materials for sodium ion batteries. *J. Mater. Chem. A* **2013**, *1* (15), 4845.
62. Su, D., Wang, G. Single-crystalline bilayered V_2O_5 nanobelts for high-capacity sodium-ion batteries. *ACS Nano* **2013**, *7* (12), 11218.

63. Dong, Y., Xu, X., Li, S., Han, C., Zhao, K., Zhang, L., Niu, C., Huang, Z., Mai, L. Inhibiting effect of Na+ pre-intercalation in MoO₃ nanobelts with enhanced electrochemical performance. *Nano Energy* **2015**, *15*, 145.
64. Wang, W., Jiang, B., Hu, L., Lin, Z., Hou, J., Jiao, S. Single crystalline VO₂ nanosheets: A cathode material for sodium-ion batteries with high rate cycling performance. *J. Power Sources* **2014**, *250*, 181.
65. Ali, G., Lee, J. H., Oh, S. H., Cho, B. W., Nam, K.-W., Chung, K. Y. Investigation of the Na intercalation mechanism into nanosized V₂O₅/C composite cathode material for Na-ion batteries. *ACS Appl. Mater. Interfaces* **2016**, *8* (9), 6032.
66. Balogun, M.-S., Luo, Y., Lyu, F., Wang, F., Yang, H., Li, H., Liang, C., Huang, M., Huang, Y., Tong, Y. Carbon quantum dot surface-engineered VO₂ interwoven nanowires: A flexible cathode material for lithium and sodium ion batteries. *ACS Appl. Mater. Interfaces* **2016**, *8* (15), 9733.
67. Chao, D., Zhu, C., Xia, X., Liu, J., Zhang, X., Wang, J., Liang, P., Lin, J., Zhang, H., Shen, Z. X. et al. Graphene quantum dots coated VO₂ arrays for highly durable electrodes for Li and Na ion batteries. *Nano Lett.* **2015**, *15* (1), 565.
68. Raju, V., Rains, J., Gates, C., Luo, W., Wang, X., Stickle, W. F., Stucky, G. D., Ji, X. Superior cathode of sodium-ion batteries: Orthorhombic V₂O₅ nanoparticles generated in nanoporous carbon by ambient hydrolysis deposition. *Nano Lett.* **2014**, *14* (7), 4119.
69. Su, D. W., Dou, S. X., Wang, G. X. Hierarchical orthorhombic V₂O₅ hollow nanospheres as high performance cathode materials for sodium-ion batteries. *J. Mater. Chem. A* **2014**, *2* (29), 11185.
70. Leong, C. C., Pan, H., Ho, S. K. Two-dimensional transition-metal oxide monolayers as cathode materials for Li and Na ion batteries. *Phys. Chem. Chem. Phys.* **2016**, *18* (10), 7527.
71. Tompsett, D. A., Islam, M. S. Electrochemistry of hollandite α-MnO₂: Li-ion and Na-ion insertion and Li₂O incorporation. *Chem. Mater.* **2013**, *25* (12), 2515.
72. Guo, S., Yu, H., Liu, D., Tian, W., Liu, X., Hanada, N., Ishida, M., Zhou, H. A novel tunnel Na$_{0.61}$Ti$_{0.48}$Mn$_{0.52}$O₂ cathode material for sodium-ion batteries. *Chem. Commun.* **2014**, *50* (59), 7998.
73. Delmas, C., Fouassier, C., Hagenmuller, P. Structural classification and properties of the layered oxides. *Physica B+C* **1980**, *99* (1), 81.
74. Doeff, M. M., Peng, M. Y., Ma, Y., De Jonghe, L. C. Orthorhombic Na$_x$MnO₂ as a cathode material for secondary sodium and lithium polymer batteries. *J. Electrochem. Soc.* **1994**, *141* (11), L145.
75. Sauvage, F., Laffont, L., Tarascon, J. M., Baudrin, E. Study of the insertion/deinsertion mechanism of sodium into Na$_{0.44}$MnO₂. *Inorg. Chem.* **2007**, *46* (8), 3289.
76. Cao, Y., Xiao, L., Wang, W., Choi, D., Nie, Z., Yu, J., Saraf, L. V., Yang, Z., Liu, J. Reversible sodium ion insertion in single crystalline manganese oxide nanowires with long cycle life. *Adv. Mater.* **2011**, *23* (28), 3155.
77. Hosono, E., Saito, T., Hoshino, J., Okubo, M., Saito, Y., Nishio-Hamane, D., Kudo, T., Zhou, H. High power Na-ion rechargeable battery with single-crystalline Na$_{0.44}$MnO₂ nanowire electrode. *J. Power Sources* **2012**, *217*, 43.
78. Kim, H., Kim, D. J., Seo, D.-H., Yeom, M. S., Kang, K., Kim, D. K., Jung, Y. Ab initio study of the sodium intercalation and intermediate phases in Na$_{0.44}$MnO₂ for sodium-ion battery. *Chem. Mater.* **2012**, *24* (6), 1205.
79. Fu, B., Zhou, X., Wang, Y. High-rate performance electrospun Na$_{0.44}$MnO₂ nanofibers as cathode material for sodium-ion batteries. *J. Power Sources* **2016**, *310*, 102.

80. Qiao, R., Dai, K., Mao, J., Weng, T.-C., Sokaras, D., Nordlund, D., Song, X., Battaglia, V. S., Hussain, Z., Liu, G. et al. Revealing and suppressing surface Mn(II) formation of Na$_{0.44}$MnO$_2$ electrodes for Na-ion batteries. *Nano Energy* **2015**, *16*, 186.
81. Cai, Y., Zhou, J., Fang, G., Cai, G., Pan, A., Liang, S. Na$_{0.282}$V$_2$O$_5$: A high-performance cathode material for rechargeable lithium batteries and sodium batteries. *J. Power Sources* **2016**, *328*, 241.
82. Jiang, X., Liu, S., Xu, H., Chen, L., Yang, J., Qian, Y. Tunnel-structured Na$_{0.54}$Mn$_{0.50}$Ti$_{0.51}$O$_2$ and Na$_{0.54}$Mn$_{0.50}$Ti$_{0.51}$O$_2$/C nanorods as advanced cathode materials for sodium-ion batteries. *Chem. Commun.* **2015**, *51* (40), 8480.
83. Yuesheng, W., Linqin, M., Jue, L., Zhenzhong, Y., Xiqian, Y., Lin, G., Yong-Sheng, H., Hong, L., Xiao-Qing, Y., Liquan, C. et al. A novel high capacity positive electrode material with tunnel-type structure for aqueous sodium-ion batteries. *Adv. Energy Mater.* **2015**, *5* (22), 1501005.
84. Wu, Z.-G., Zhong, Y.-J., Li, J.-T., Wang, K., Guo, X.-D., Huang, L., Zhong, B.-H., Sun, S.-G. Synthesis of a novel tunnel Na$_{0.5}$K$_{0.1}$MnO$_2$ composite as a cathode for sodium ion batteries. *RSC Adv.* **2016**, *6* (59), 54404.
85. Kuhn, A., Menéndez, N., García-Alvarado, F., Morán, E., Tornero, J. D., Alario-Franco, M. A. Topotactic oxidation of the quadruple-rutile-type chain structure Na$_{0.875}$Fe$_{0.875}$Ti$_{1.125}$O$_4$. *J. Solid State Chem.* **1997**, *130* (2), 184.
86. Shuyin, X., Yuesheng, W., Liubin, B., Yingchun, L., Ningning, S., Zhenzhong, Y., Yunming, L., Linqin, M., Hai-Tao, Y., Lin, G. et al. Fe-based tunnel-type Na$_{0.61}$[Mn$_{0.27}$Fe$_{0.34}$Ti$_{0.39}$]O$_2$ designed by a new strategy as a cathode material for sodium-ion batteries. *Adv. Energy Mater.* **2015**, *5* (22), 1501156.
87. Zhang, L., Ji, S., Yu, L., Xu, X., Liu, J. Amorphous FeF3/C nanocomposite cathode derived from metal–organic frameworks for sodium ion batteries. *RSC Adv.* **2017**, *7* (39), 24004.
88. Ma, D.-L., Wang, H.-G., Li, Y., Xu, D., Yuan, S., Huang, X.-L., Zhang, X.-B., Zhang, Y. In situ generated FeF$_3$ in homogeneous iron matrix toward high-performance cathode material for sodium-ion batteries. *Nano Energy* **2014**, *10*, 295.
89. Zhou, Y. N., Sina, M., Pereira, N., Yu, X., Amatucci, G. G., Yang, X. Q., Cosandey, F., Nam, K. W. FeO$_{0.7}$F$_{1.3}$/C nanocomposite as a high-capacity cathode material for sodium-ion batteries. *Adv. Funct. Mater.* **2015**, *25* (5), 696.
90. Dimov, N., Nishimura, A., Chihara, K., Kitajou, A., Gocheva, I. D., Okada, S. Transition metal NaMF$_3$ compounds as model systems for studying the feasibility of ternary Li-M-F and Na-M-F single phases as cathodes for lithium-ion and sodium-ion batteries. *Electrochim. Acta* **2013**, *110*, 214.
91. Nava-Avendaño, J., Arroyo-de Dompablo, M. E., Frontera, C., Ayllón, J. A., Palacín, M. R. Study of sodium manganese fluorides as positive electrodes for Na-ion batteries. *Solid State Ion.* **2015**, *278*, 106.
92. Gocheva, I. D., Nishijima, M., Doi, T., Okada, S., Yamaki, J.-I., Nishida, T. Mechanochemical synthesis of NaMF$_3$ (M=Fe, Mn, Ni) and their electrochemical properties as positive electrode materials for sodium batteries. *J. Power Sources* **2009**, *187* (1), 247.
93. Yamada, Y., Doi, T., Tanaka, I., Okada, S., Yamaki, J.-I. Liquid-phase synthesis of highly dispersed NaFeF$_3$ particles and their electrochemical properties for sodium-ion batteries. *J. Power Sources* **2011**, *196* (10), 4837.

94. Kitajou, A., Komatsu, H., Chihara, K., Gocheva, I. D., Okada, S., Yamaki, J.-I. Novel synthesis and electrochemical properties of perovskite-type NaFeF$_3$ for a sodium-ion battery. *J. Power Sources* **2012**, *198*, 389.
95. Yin, X., Huang, K., Liu, S., Wang, H., Wang, H. Preparation and characterization of Na-doped LiFePO$_4$/C composites as cathode materials for lithium-ion batteries. *J. Power Sources* **2010**, *195* (13), 4308.
96. Wang, J., Sun, X. Olivine LiFePO$_4$: The remaining challenges for future energy storage. *Energy Environ. Sci.* **2015**, *8* (4), 1110.
97. Essehli, R., Ben Yahia, H., Maher, K., Sougrati, M. T., Abouimrane, A., Park, J. B., Sun, Y. K., Al-Maadeed, M. A., Belharouak, I. Unveiling the sodium intercalation properties in Na$_{1.860.14}$Fe$_3$(PO$_4$)$_3$. *J. Power Sources* **2016**, *324*, 657.
98. Moreau, P., Guyomard, D., Gaubicher, J., Boucher, F. Structure and stability of sodium intercalated phases in olivine FePO$_4$. *Chem. Mater.* **2010**, *22* (14), 4126.
99. Shiratsuchi, T., Okada, S., Yamaki, J., Nishida, T. FePO$_4$ cathode properties for Li and Na secondary cells. *J. Power Sources* **2006**, *159* (1), 268.
100. Junmei, Z., Zelang, J., Jie, M., Fuchun, W., Yong-Sheng, H., Wen, C., Liquan, C., Huizhou, L., Sheng, D. Monodisperse iron phosphate nanospheres: Preparation and application in energy storage. *ChemSusChem* **2012**, *5* (8), 1495.
101. Liu, Y., Xu, Y., Han, X., Pellegrinelli, C., Zhu, Y., Zhu, H., Wan, J., Chung, A. C., Vaaland, O., Wang, C. et al. Porous amorphous FePO$_4$ nanoparticles connected by single-wall carbon nanotubes for sodium ion battery cathodes. *Nano Lett.* **2012**, *12* (11), 5664.
102. Bridson, J. N., Quinlan, S. E., Tremaine, P. R. Synthesis and crystal structure of maricite and sodium iron(III) hydroxyphosphate. *Chem. Mater.* **1998**, *10* (3), 763.
103. Hammond, R., Barbier, J. Structural chemistry of NaCoPO$_4$. *Acta Crystallographica Section B* **1996**, *52* (3), 440.
104. Zaghib, K., Trottier, J., Hovington, P., Brochu, F., Guerfi, A., Mauger, A., Julien, C. M. Characterization of Na-based phosphate as electrode materials for electrochemical cells. *J. Power Sources* **2011**, *196* (22), 9612.
105. Moring, J., Kostiner, E. The crystal structure of NaMnPO$_4$. *J. Solid State Chem.* **1986**, *61* (3), 379.
106. Fernández-Ropero, A. J., Saurel, D., Acebedo, B., Rojo, T., Casas-Cabanas, M. Electrochemical characterization of NaFePO$_4$ as positive electrode in aqueous sodium-ion batteries. *J. Power Sources* **2015**, *291*, 40.
107. Lee, K. T., Ramesh, T. N., Nan, F., Botton, G., Nazar, L. F. Topochemical synthesis of sodium metal phosphate olivines for sodium-ion batteries. *Chem. Mater.* **2011**, *23* (16), 3593.
108. Tealdi, C., Heath, J., Islam, M. S. Feeling the strain: Enhancing ionic transport in olivine phosphate cathodes for Li- and Na-ion batteries through strain effects. *J. Mater. Chem. A* **2016**, *4* (18), 6998.
109. Ali, G., Lee, J.-H., Susanto, D., Choi, S.-W., Cho, B. W., Nam, K.-W., Chung, K. Y. Polythiophene-wrapped olivine NaFePO$_4$ as a cathode for Na-Ion batteries. *ACS Appl. Mater. Interfaces* **2016**, *8* (24), 15422.
110. Trad, K., Carlier, D., Croguennec, L., Wattiaux, A., Ben Amara, M., Delmas, C. NaMnFe$_2$(PO$_4$)$_3$ Alluaudite phase: Synthesis, structure, and electrochemical properties as positive electrode in lithium and sodium batteries. *Chem. Mater.* **2010**, *22* (19), 5554.

111. Kim, J., Seo, D.-H., Kim, H., Park, I., Yoo, J.-K., Jung, S.-K., Park, Y.-U., Goddard III, W. A., Kang, K. Unexpected discovery of low-cost maricite NaFePO$_4$ as a high-performance electrode for Na-ion batteries. *Energy Environ. Sci.* **2015**, *8* (2), 540.

112. Yang, G., Ding, B., Wang, J., Nie, P., Dou, H., Zhang, X. Excellent cycling stability and superior rate capability of a graphene–amorphous FePO$_4$ porous nanowire hybrid as a cathode material for sodium ion batteries. *Nanoscale* **2016**, *8* (16), 8495.

113. Fang, Y., Xiao, L., Qian, J., Ai, X., Yang, H., Cao, Y. Mesoporous amorphous FePO$_4$ nanospheres as high-performance cathode material for sodium-ion batteries. *Nano Lett.* **2014**, *14* (6), 3539.

114. Ramireddy, T., Rahman, M. M., Sharma, N., Glushenkov, A. M., Chen, Y. Carbon coated Na$_7$Fe$_7$(PO$_4$)$_6$F$_3$: A novel intercalation cathode for sodium-ion batteries. *J. Power Sources* **2014**, *271*, 497.

115. Tripathi, R., Wood, S. M., Islam, M. S., Nazar, L. F. Na-ion mobility in layered Na$_2$FePO$_4$F and olivine Na[Fe,Mn]PO$_4$. *Energy Environ. Sci.* **2013**, *6* (8), 2257.

116. Ong, S. P., Chevrier, V. L., Ceder, G. Comparison of small polaron migration and phase separation in olivine LiMnPO$_4$ and LiFePO$_4$ using hybrid density functional theory. *Phys. Rev. B* **2011**, *83* (7), 075112.

117. Hong, H. Y. P. Crystal structures and crystal chemistry in the system Na$_{1+x}$Zr$_2$Si$_x$P$_3$–xO$_{12}$. *Mater. Res. Bull.* **1976**, *11* (2), 173.

118. Goodenough, J. B., Hong, H. Y. P., Kafalas, J. A. Fast Na+-ion transport in skeleton structures. *Mater. Res. Bull.* **1976**, *11* (2), 203.

119. Zatovsky, I. V. NASICON-type Na$_3$V$_2$(PO$_4$)$_3$. *Acta Crystallographica Section E: Structure Reports Online* **2010**, *66* (Pt 2), i12.

120. Jian, Z., Zhao, L., Pan, H., Hu, Y.-S., Li, H., Chen, W., Chen, L. Carbon coated Na$_3$V$_2$(PO$_4$)$_3$ as novel electrode material for sodium ion batteries. *Electrochem. Commun.* **2012**, *14* (1), 86.

121. Ranjusha, R., Lei, Z., Xue, D. S., Kun, L. H. Tuned in situ growth of nanolayered rGO on 3D Na$_3$V$_2$(PO$_4$)$_3$ matrices: A step toward long lasting, high power Na-ion batteries. *Adv. Mater. Interfaces* **2016**, *3* (13), 1600007.

122. Yasushi, U., Toshiyasu, K., Shigeto, O., Jun-Ichi, Y. Electrochemical sodium insertion into the 3D-framework of Na$_3$M$_2$(PO$_4$)$_3$ (M=Fe, V). *The Reports of Institute of Advanced Material Study Kyushu University* **2002**, *16*, 5.

123. Plashnitsa, L. S., Kobayashi, E., Noguchi, Y., Okada, S., Yamaki, J.-I. Performance of NASICON symmetric cell with ionic liquid electrolyte. *J. Electrochem. Soc.* **2010**, *157* (4), A536.

124. Song, W., Ji, X., Yao, Y., Zhu, H., Chen, Q., Sun, Q., Banks, C. E. A promising Na$_3$V$_2$(PO$_4$)$_3$ cathode for use in the construction of high energy batteries. *Phys. Chem. Chem. Phys. : PCCP* **2014**, *16* (7), 3055.

125. Shen, W., Wang, C., Liu, H., Yang, W. Towards highly stable storage of sodium ions: A porous Na$_3$V$_2$(PO$_4$)$_3$/C cathode material for sodium-ion batteries. *Chemistry* **2013**, *19* (43), 14712.

126. Qian, J., Zhou, M., Cao, Y., Ai, X., Yang, H. Nanosized Na$_4$Fe(CN)$_6$/C composite as a low-cost and high-rate cathode material for sodium-ion batteries. *Adv. Energy Mater.* **2012**, *2* (4), 410.

127. Ren, M., Zhou, Z., Li, Y., Gao, X. P., Yan, J. Preparation and electrochemical studies of Fe-doped Li$_3$V$_2$(PO$_4$)$_3$ cathode materials for lithium-ion batteries. *J. Power Sources* **2006**, *162* (2), 1357.
128. Wang, Q., Zhao, B., Zhang, S., Gao, X., Deng, C. Superior sodium intercalation of honeycomb-structured hierarchical porous Na$_3$V$_2$(PO$_4$)$_3$/C microballs prepared by a facile one-pot synthesis. *J. Mater. Chem. A* **2015**, *3* (15), 7732.
129. Saravanan, K., Mason, C. W., Rudola, A., Wong, K. H., Balaya, P. The first report on excellent cycling stability and superior rate capability of Na$_3$V$_2$(PO$_4$)$_3$ for sodium ion batteries. *Adv. Energy Mater.* **2013**, *3* (4), 444.
130. Li, H., Yu, X., Bai, Y., Wu, F., Wu, C., Liu, L., Yang, X.-Q. Effects of Mg doping on remarkably enhanced electrochemistry performances of Na$_3$V$_2$(PO$_4$)$_3$ cathode material for sodium ion batteries. *J. Mater. Chem.* **2015**, *3*, 9578.
131. Klee, R., Lavela, P., Aragón, M. J., Alcántara, R., Tirado, J. L. Enhanced high-rate performance of manganese substituted Na$_3$V$_2$(PO$_4$)$_3$/C as cathode for sodium-ion batteries. *J. Power Sources* **2016**, *313*, 73.
132. Ranjusha, R., Bo, C., Zhicheng, Z., Xing-Long, W., Yonghua, D., Ying, H., Bing, L., Yun, Z., Jie, W., Gwang-Hyeon, N. et al. Improved reversibility of Fe^{3+}/Fe^{4+} redox couple in sodium super ion conductor type Na$_3$Fe$_2$(PO$_4$)$_3$ for sodium-ion batteries. *Adv. Mater.* **2017**, *29* (12), 1605694.
133. Song, J., Xu, M., Wang, L., Goodenough, J. B. Exploration of NaVOPO$_4$ as a cathode for a Na-ion battery. *Chem. Commun.* **2013**, *49* (46), 5280.
134. He, G., Huq, A., Kan, W. H., Manthiram, A. β-NaVOPO$_4$ obtained by a low-temperature synthesis process: A new 3.3 V cathode for sodium-ion batteries. *Chem. Mater.* **2016**, *28* (5), 1503.
135. He, G., Kan, W. H., Manthiram, A. A 3.4 V layered VOPO$_4$ cathode for Na-ion batteries. *Chem. Mater.* **2016**, *28* (2), 682.
136. Kundu, D., Tripathi, R., Popov, G., Makahnouk, W. R. M., Nazar, L. F. Synthesis, structure, and Na-ion migration in Na$_4$NiP$_2$O$_7$F$_2$: A prospective high voltage positive electrode material for the Na-ion battery. *Chem. Mater.* **2015**, *27* (3), 885.
137. Jongsoon, K., Inchul, P., Hyungsub, K., Kyu-Young, P., Young-Uk, P., Kisuk, K. Tailoring a new 4V-class cathode material for Na-ion batteries. *Adv. Energy Mater.* **2016**, *6* (6), 1502147.
138. Barpanda, P., Avdeev, M., Ling, C. D., Lu, J., Yamada, A. Magnetic structure and properties of the Na$_2$CoP$_2$O$_7$ pyrophosphate cathode for sodium-ion batteries: A supersuperexchange-driven non-collinear antiferromagnet. *Inorg. Chem.* **2013**, *52* (1), 395.
139. Barpanda, P., Liu, G., Ling, C. D., Tamaru, M., Avdeev, M., Chung, S.-C., Yamada, Y., Yamada, A. Na$_2$FeP$_2$O$_7$: A safe cathode for rechargeable sodium-ion batteries. *Chem. Mater.* **2013**, *25* (17), 3480.
140. Barpanda, P., Ye, T., Nishimura, S.-I., Chung, S.-C., Yamada, Y., Okubo, M., Zhou, H., Yamada, A. Sodium iron pyrophosphate: A novel 3.0 V iron-based cathode for sodium-ion batteries. *Electrochem. Commun.* **2012**, *24*, 116.
141. Heejin, K., Sun, P. C., Wook, C. J., Yousung, J. Defect-controlled formation of triclinic Na$_2$CoP$_2$O$_7$ for 4 V sodium-ion batteries. *Angewandte Chemie International Edition* **2016**, *55* (23), 6662.

142. Kim, H., Shakoor, R. A., Park, C., Lim, S. Y., Kim, J. S., Jo, Y. N., Cho, W., Miyasaka, K., Kahraman, R., Jung, Y. et al. Na$_2$FeP$_2$O$_7$ as a promising iron-based pyrophosphate cathode for sodium rechargeable batteries: A combined experimental and theoretical study. *Adv. Funct. Mater.* **2013**, *23* (9), 1147.

143. Longoni, G., Wang, J. E., Jung, Y. H., Kim, D. K., Mari, C. M., Ruffo, R. The Na$_2$FeP$_2$O$_7$-carbon nanotubes composite as high rate cathode material for sodium ion batteries. *J. Power Sources* **2016**, *302*, 61.

144. Lin, B., Li, Q., Liu, B., Zhang, S., Deng, C. Biochemistry-directed hollow porous microspheres: Bottom-up self-assembled polyanion-based cathodes for sodium ion batteries. *Nanoscale* **2016**, *8* (15), 8178.

145. Niu, Y., Xu, M., Cheng, C., Bao, S., Hou, J., Liu, S., Yi, F., He, H., Li, C. M. Na$_{3.12}$Fe$_{2.44}$(P$_2$O$_7$)$_2$/multi-walled carbon nanotube composite as a cathode material for sodium-ion batteries. *J. Mater. Chem. A* **2015**, *3* (33), 17224.

146. Barpanda, P., Ye, T., Avdeev, M., Chung, S.-C., Yamada, A. A new polymorph of Na$_2$MnP$_2$O$_7$ as a 3.6 V cathode material for sodium-ion batteries. *J. Mater. Chem. A* **2013**, *1* (13), 4194.

147. Park, C. S., Kim, H., Shakoor, R. A., Yang, E., Lim, S. Y., Kahraman, R., Jung, Y., Choi, J. W. Anomalous manganese activation of a pyrophosphate cathode in sodium ion batteries: A combined experimental and theoretical study. *J. Am. Chem. Soc.* **2013**, *135* (7), 2787.

148. Barpanda, P., Lu, J., Ye, T., Kajiyama, M., Chung, S.-C., Yabuuchi, N., Komaba, S., Yamada, A. A layer-structured Na$_2$CoP$_2$O$_7$ pyrophosphate cathode for sodium-ion batteries. *RSC Adv.* **2013**, *3* (12), 3857.

149. Ha, K. H., Woo, S. H., Mok, D., Choi, N.-S., Park, Y., Oh, S. M., Kim, Y., Kim, J., Lee, J., Nazar, L. F. et al. Na$_4$-αM$_2$+α/$_2$(P$_2$O$_7$)$_2$ (2/3 ≤ α ≤ 7/8, M = Fe, Fe$_{0.5}$Mn$_{0.5}$, Mn): A promising sodium ion cathode for Na-ion batteries. *Adv. Energy Mater.* **2013**, *3* (6), 770.

150. Shakoor, R. A., Park, C. S., Raja, A. A., Shin, J., Kahraman, R. A mixed iron–manganese based pyrophosphate cathode, Na$_2$Fe$_{0.5}$Mn$_{0.5}$P$_2$O$_7$, for rechargeable sodium ion batteries. *Phys. Chem. Chem. Phys.* **2016**, *18* (5), 3929.

151. Barker, J., Gover, R. K. B., Burns, P., Bryan, A. A symmetrical lithium-ion cell based on lithium vanadium fluorophosphate, LiVPO$_4$F. *Electrochem. Solid-State Lett.* **2005**, *8* (6), A285.

152. Kawabe, Y., Yabuuchi, N., Kajiyama, M., Fukuhara, N., Inamasu, T., Okuyama, R., Nakai, I., Komaba, S. A comparison of crystal structures and electrode performance between Na$_2$FePO$_4$F and Na$_2$Fe$_{0.5}$Mn$_{0.5}$PO$_4$F synthesized by solid-state method for rechargeable Na-ion batteries. *Electrochemistry* **2012**, *80* (2), 80.

153. Chihara, K., Kitajou, A., Gocheva, I. D., Okada, S., Yamaki, J.-I. Cathode properties of Na$_3$M$_2$(PO$_4$)$_2$F$_3$ [M = Ti, Fe, V] for sodium-ion batteries. *J. Power Sources* **2013**, *227*, 80.

154. Langrock, A., Xu, Y., Liu, Y., Ehrman, S., Manivannan, A., Wang, C. Carbon coated hollow Na$_2$FePO$_4$F spheres for Na-ion battery cathodes. *J. Power Sources* **2013**, *223*, 62.

155. Xu, M., Wang, L., Zhao, X., Song, J., Xie, H., Lu, Y., Goodenough, J. B. Na$_3$V$_2$O$_2$(PO$_4$)$_2$F/graphene sandwich structure for high-performance cathode of a sodium-ion battery. *Phys. Chem. Chem. Phys.* **2013**, *15* (31), 13032.

156. Law, M., Balaya, P. NaVPO$_4$F with high cycling stability as a promising cathode for sodium-ion battery. *Energy Storage Materials* **2018**, *10*, 102.

157. Shakoor, R. A., Seo, D.-H., Kim, H., Park, Y.-U., Kim, J., Kim, S.-W., Gwon, H., Lee, S., Kang, K. A combined first principles and experimental study on $Na_3V_2(PO_4)_2F_3$ for rechargeable Na batteries. *J. Mater. Chem.* **2012**, *22* (38), 20535.
158. Song, W., Ji, X., Wu, Z., Yang, Y., Zhou, Z., Li, F., Chen, Q., Banks, C. E. Exploration of ion migration mechanism and diffusion capability for $Na_3V_2(PO_4)_2F_3$ cathode utilized in rechargeable sodium-ion batteries. *J. Power Sources* **2014**, *256*, 258.
159. Bianchini, M., Brisset, N., Fauth, F., Weill, F., Elkaim, E., Suard, E., Masquelier, C., Croguennec, L. $Na_3V_2(PO_4)_2F_3$ revisited: A high-resolution diffraction study. *Chem. Mater.* **2014**, *26* (14), 4238.
160. Barker, J., Saidi, M. Y., Swoyer, J. L. A sodium-ion cell based on the fluorophosphate compound $NaVPO_4F$. *Electrochem. Solid-State Lett.* **2003**, *6* (1), A1.
161. Barker, J., Saidi, M. Y., Swoyer, J. L. A comparative investigation of the Li insertion properties of the novel fluorophosphate phases, $NaVPO_4F$ and $LiVPO_4F$. *J. Electrochem. Soc.* **2004**, *151* (10), A1670.
162. Zhuo, H., Wang, X., Tang, A., Liu, Z., Gamboa, S., Sebastian, P. J. The preparation of $NaV_{1-x}Cr_xPO_4F$ cathode materials for sodium-ion battery. *J. Power Sources* **2006**, *160* (1), 698.
163. Matts, I. L., Dacek, S., Pietrzak, T. K., Malik, R., Ceder, G. Explaining performance-limiting mechanisms in fluorophosphate Na-ion battery cathodes through inactive transition-metal mixing and first-principles mobility calculations. *Chem. Mater.* **2015**, *27* (17), 6008.
164. Serras, P., Palomares, V., Goñi, A., Gil de Muro, I., Kubiak, P., Lezama, L., Rojo, T. High voltage cathode materials for Na-ion batteries of general formula $Na_3V_2O_{2x}(PO_4)_2F_{3-2x}$. *J. Mater. Chem.* **2012**, *22* (41), 22301.
165. Serras, P., Palomares, V., Goñi, A., Kubiak, P., Rojo, T. Electrochemical performance of mixed valence $Na_3V_2O_{2x}(PO_4)_2F_{3-2x}$/C as cathode for sodium-ion batteries. *J. Power Sources* **2013**, *241*, 56.
166. Sharma, N., Serras, P., Palomares, V., Brand, H. E. A., Alonso, J., Kubiak, P., Fdez-Gubieda, M. L., Rojo, T. Sodium distribution and reaction mechanisms of a $Na_3V_2O_2(PO_4)_2F$ electrode during use in a sodium-ion battery. *Chem. Mater.* **2014**, *26* (11), 3391.
167. Manhua, P., Biao, L., Huijun, Y., Dongtang, Z., Xiayan, W., Dingguo, X., Guangsheng, G. Ruthenium-oxide-coated sodium vanadium fluorophosphate nanowires as high-power cathode materials for sodium-ion batteries. *Angewandte Chemie International Edition* **2015**, *54* (22), 6452.
168. Kumar, P. R., Jung, Y. H., Lim, C. H., Kim, D. K. $Na_3V_2O_{2x}(PO_4)_2F_{3-2x}$: A stable and high-voltage cathode material for aqueous sodium-ion batteries with high energy density. *J. Mater. Chem. A* **2015**, *3* (12), 6271.
169. Serras, P., Palomares, V., Alonso, J., Sharma, N., López del Amo, J. M., Kubiak, P., Fdez-Gubieda, M. L., Rojo, T. Electrochemical Na extraction/insertion of $Na_3V_2O_{2x}(PO_4)_2F_{3-2x}$. *Chem. Mater.* **2013**, *25* (24), 4917.
170. Qi, Y., Mu, L., Zhao, J., Hu, Y.-S., Liu, H., Dai, S. Superior Na-storage performance of low-temperature-synthesized $Na_3(VO_{1-x}PO_4)_2F_{1+2x}$ (0≤x≤1) nanoparticles for Na-ion batteries. *Angewandte Chemie International Edition* **2015**, *54* (34), 9911.
171. Park, Y.-U., Seo, D.-H., Kwon, H.-S., Kim, B., Kim, J., Kim, H., Kim, I., Yoo, H.-I., Kang, K. A new high-energy cathode for a Na-ion battery with ultrahigh stability. *J. Am. Chem. Soc.* **2013**, *135* (37), 13870.

172. Zhang, B., Dugas, R., Rousse, G., Rozier, P., Abakumov, A. M., Tarascon, J.-M. Insertion compounds and composites made by ball milling for advanced sodium-ion batteries. *Nat. Commun.* **2016**, *7*, 10308.
173. Ellis, B. L., Makahnouk, W. R. M., Makimura, Y., Toghill, K., Nazar, L. F. A multifunctional 3.5 V iron-based phosphate cathode for rechargeable batteries. *Nat. Mater.* **2007**, *6*, 749.
174. Kim, S.-W., Seo, D.-H., Kim, H., Park, K.-Y., Kang, K. A comparative study on Na_2MnPO_4F and Li_2MnPO_4F for rechargeable battery cathodes. *Phys. Chem. Chem. Phys.* **2012**, *14* (10), 3299.
175. Kim, H., Park, I., Seo, D.-H., Lee, S., Kim, S.-W., Kwon, W. J., Park, Y.-U., Kim, C. S., Jeon, S., Kang, K. New iron-based mixed-polyanion cathodes for lithium and sodium rechargeable batteries: Combined first principles calculations and experimental study. *J. Am. Chem. Soc.* **2012**, *134* (25), 10369.
176. Kim, H., Park, I., Lee, S., Kim, H., Park, K.-Y., Park, Y.-U., Kim, H., Kim, J., Lim, H.-D., Yoon, W.-S. et al. Understanding the electrochemical mechanism of the new iron-based mixed-phosphate $Na_4Fe_3(PO_4)_2(P_2O_7)$ in a Na rechargeable battery. *Chem. Mater.* **2013**, *25* (18), 3614.
177. Lee, Y., Lee, J., Kim, H., Kang, K., Choi, N.-S. Highly stable linear carbonate-containing electrolytes with fluoroethylene carbonate for high-performance cathodes in sodium-ion batteries. *J. Power Sources* **2016**, *320*, 49.
178. Sanz, F., Parada, C., Rojo, J. M., Ruíz-Valero, C. Synthesis, structural characterization, magnetic properties, and ionic conductivity of $Na_4MII_3(PO_4)_2(P_2O_7)$ (MII = Mn, Co, Ni). *Chem. Mater.* **2001**, *13* (4), 1334.
179. Kim, H., Yoon, G., Park, I., Park, K.-Y., Lee, B., Kim, J., Park, Y.-U., Jung, S.-K., Lim, H.-D., Ahn, D. et al. Anomalous Jahn–Teller behavior in a manganese-based mixed-phosphate cathode for sodium ion batteries. *Energy Environ. Sci.* **2015**, *8* (11), 3325.
180. Deng, C., Zhang, S. 1D nanostructured $Na_7V_4(P_2O_7)_4(PO_4)$ as high-potential and superior-performance cathode material for sodium-ion batteries. *ACS Appl. Mater. Interfaces* **2014**, *6* (12), 9111.
181. Hassanzadeh, N., Sadrnezhaad, S. K., Chen, G. In-situ hydrothermal synthesis of $Na_3MnCO_3PO_4$/rGO hybrid as a cathode for Na-ion battery. *Electrochim. Acta* **2016**, *208*, 188.
182. Chen, H., Hao, Q., Zivkovic, O., Hautier, G., Du, L.-S., Tang, Y., Hu, Y.-Y., Ma, X., Grey, C. P., Ceder, G. Sidorenkite ($Na_3MnPO_4CO_3$): A new intercalation cathode material for Na-ion batteries. *Chem. Mater.* **2013**, *25* (14), 2777.
183. Moore, P. B. Crystal chemistry of the alluaudite structure type: Contribution to the paragenesis of pegmatite phosphate giant crystals. *Am. Mineral.* **1971**, *56* (11–12), 1955.
184. Palomares, V., Serras, P., Villaluenga, I., Hueso, K. B., Carretero-González, J., Rojo, T. Na-ion batteries, recent advances and present challenges to become low cost energy storage systems. *Energy Environ. Sci.* **2012**, *5* (3), 5884.
185. Palmore, G. T. R., Liu, D. A new cathode for sodium ion batteries: Alluaudite $Na_{1.702}Fe_3(PO_4)_3$. *Meeting Abstracts* **2016**, *MA2016-01* (3), 393.
186. Wen, M., Liu, U., Zhao, Y., Liu, S., Liu, H., Dong, Y., Kuang, Q., Fan, Q. Synthesis of alluaudite-type $Na_2VFe_2(PO_4)_3$/C and its electrochemical performance as cathode material for sodium-ion battery. *J. Solid State Electrochem.* **2018**, *22* (3), 8.

187. Jiangfeng, Q., Chen, W., Yuliang, C., Zifeng, M., Yunhui, H., Xinping, A., Hanxi, Y. Prussian blue cathode materials for sodium-ion batteries and other ion batteries. *Adv. Energy Mater.* **2018**, *8* (17), 1702619.
188. Wessells, C. D., Peddada, S. V., Huggins, R. A., Cui, Y. Nickel hexacyanoferrate nanoparticle electrodes for aqueous sodium and potassium ion batteries. *Nano Lett.* **2011**, *11* (12), 5421.
189. Pasta, M., Wessells, C. D., Huggins, R. A., Cui, Y. A high-rate and long cycle life aqueous electrolyte battery for grid-scale energy storage. *Nat. Commun.* **2012**, *3*, 1149.
190. Lu, Y., Wang, L., Cheng, J., Goodenough, J. B. Prussian blue: A new framework of electrode materials for sodium batteries. *Chem. Commun.* **2012**, *48* (52), 6544.
191. Wang, L., Lu, Y., Liu, J., Xu, M., Cheng, J., Zhang, D., Goodenough., J. B. A superior low-cost cathode for a Na-ion battery. *Angewandte Chemie International Edition* **2013**, *52* (7), 1964.
192. Jianfeng, Q., Min, Z., Yuliang, C., Xinping, A., Hanxi, Y. Nanosized Na$_4$Fe(CN)$_6$/C composite as a low-cost and high-rate cathode material for sodium-ion batteries. *Adv. Energy Mater.* **2012**, *2* (4), 410.
193. Novák, P., Müller, K., Santhanam, K. S. V., Haas, O. Electrochemically active polymers for rechargeable batteries. *Chemical Reviews* **1997**, *97* (1), 207.
194. Wang, L., Song, J., Qiao, R., Wray, L. A., Hossain, M. A., Chuang, Y.-D., Yang, W., Lu, Y., Evans, D., Lee, J.-J. et al. Rhombohedral Prussian white as cathode for rechargeable sodium-ion batteries. *J. Am. Chem. Soc.* **2015**, *137* (7), 2548.
195. Lee, H.-W., Wang, R. Y., Pasta, M., Woo Lee, S., Liu, N., Cui, Y. Manganese hexacyanomanganate open framework as a high-capacity positive electrode material for sodium-ion batteries. *Nat. Commun.* **2014**, *5*, 5280.
196. Fu, H., Liu, C., Zhang, C., Ma, W., Wang, K., Li, Z., Lu, X., Cao, G. Enhanced storage of sodium ions in Prussian blue cathode material through nickel doping. *J. Mater. Chem. A* **2017**, *5* (20), 9604.
197. Yang, D., Xu, J., Liao, X.-Z., He, Y.-S., Liu, H., Ma, Z.-F. Structure optimization of Prussian blue analogue cathode materials for advanced sodium ion batteries. *Chem. Commun.* **2014**, *50* (87), 13377.
198. Padhi, A. K., Nanjundaswamy, K. S., Masquelier, C., Goodenough, J. B. Mapping of transition metal redox energies in phosphates with NASICON structure by lithium intercalation. *J. Electrochem. Soc.* **1997**, *144* (8), 2581.
199. Barpanda, P., Oyama, G., Nishimura, S.-I., Chung, S.-C., Yamada, A. A 3.8-V earth-abundant sodium battery electrode. *Nat. Commun.* **2014**, *5*, 4358.
200. Barpanda, P., Oyama, G., Ling, C. D., Yamada, A. Kröhnkite-type Na$_2$Fe(SO$_4$)$_2$·2H$_2$O as a novel 3.25 V insertion compound for Na-ion batteries. *Chem. Mater.* **2014**, *26* (3), 1297.
201. Meng, Y., Zhang, S., Deng, C. Superior sodium–lithium intercalation and depressed moisture sensitivity of a hierarchical sandwich-type nanostructure for a graphene–sulfate composite: A case study on Na$_2$Fe(SO$_4$)$_2$·2H$_2$O. *J. Mater. Chem. A* **2015**, *3* (8), 4484.
202. Mason, C. W., Gocheva, I., Hoster, H. E., Yu, D. Y. W. Iron(iii) sulfate: A stable, cost effective electrode material for sodium ion batteries. *Chem. Commun.* **2014**, *50* (18), 2249.
203. Singh, P., Shiva, K., Celio, H., Goodenough, J. B. Eldfellite, NaFe(SO$_4$)$_2$: An intercalation cathode host for low-cost Na-ion batteries. *Energy Environ. Sci.* **2015**, *8* (10), 3000.

204. Suhao, W., Benoit, M. D. B., Gyosuke, O., Shin-ichi, N., Atsuo, Y. Synthesis and electrochemistry of $Na_{2.5}(Fe_{1-y}Mn_y)1.75(SO_4)_3$ solid solutions for Na-ion batteries. *ChemElectroChem* **2016**, *3* (2), 209.
205. Barpanda, P., Chotard, J.-N., Recham, N., Delacourt, C., Ati, M., Dupont, L., Armand, M., Tarascon, J.-M. Structural, transport, and electrochemical investigation of novel $AMSO_4F$ (A = Na, Li, M = Fe, Co, Ni, Mn) metal fluorosulphates prepared using low temperature synthesis routes. *Inorg. Chem.* **2010**, *49* (16), 7401.
206. Reynaud, M., Barpanda, P., Rousse, G., Chotard, J.-N., Melot, B. C., Recham, N., Tarascon, J.-M. Synthesis and crystal chemistry of the $NaMSO_4F$ family (M = Mg, Fe, Co, Cu, Zn). *Solid State Sci.* **2012**, *14* (1), 15.
207. Tripathi, R., Ramesh, T. N., Ellis, B. L., Nazar, L. F. Scalable synthesis of tavorite $LiFeSO_4F$ and $NaFeSO_4F$ cathode materials. *Angewandte Chemie International Edition* **2010**, *49* (46), 8738.
208. Tripathi, R., Gardiner, G. R., Islam, M. S., Nazar, L. F. Alkali-ion conduction paths in $LiFeSO_4F$ and $NaFeSO_4F$ tavorite-type cathode materials. *Chem. Mater.* **2011**, *23* (8), 2278.
209. Melot, B. C., Rousse, G., Chotard, J.-N., Kemei, M. C., Rodriguez-Carvajal, J., Tarascon, J. M. Magnetic structure and properties of $NaFeSO_4F$ and $NaCoSO_4F$. *Phys. Rev. B* **2012**, *85*, 094415.
210. Rajagopalan, R., Wu, Z., Liu, Y., Al-Rubaye, S., Wang, E., Wu, C., Xiang, W., Zhong, B., Guo, X., Dou, S. X. et al. A novel high voltage battery cathodes of Fe^{2+}/Fe^{3+} sodium fluoro sulfate lined with carbon nanotubes for stable sodium batteries. *J. Power Sources* **2018**, *398*, 175.
211. Haiyan, C., Michel, A., Gilles, D., Franck, D., Philippe, P., Jean-Marie, T. From biomass to a renewable $LiXC_6O_6$ organic electrode for sustainable Li-ion batteries. *ChemSusChem* **2008**, *1* (4), 348.
212. Bostjan, G., Klemen, P., Romana, C.-K., Robert, D., Miran, G. Electroactive organic molecules immobilized onto solid nanoparticles as a cathode material for lithium-ion batteries. *Angewandte Chemie International Edition* **2010**, *49* (40), 7222.
213. Ratnakumar, B. V., Di Stefano, S., Williams, R. M., Nagasubramanian, G., Bankston, C. P. Organic cathode materials in sodium batteries. *J. Appl. Electrochem.* **1990**, *20* (3), 357.
214. Chihara, K., Chujo, N., Kitajou, A., Okada, S. Cathode properties of $Na_2C_6O_6$ for sodium-ion batteries. *Electrochim. Acta* **2013**, *110*, 240.
215. Wang, Y., Ding, Y., Pan, L., Shi, Y., Yue, Z., Shi, Y., Yu, G. Understanding the size-dependent sodium storage properties of $Na_2C_6O_6$-based organic electrodes for sodium-ion batteries. *Nano Lett.* **2016**, *16* (5), 3329.
216. Chengliang, W., Yaoguo, F., Yang, X., Liying, L., Min, Z., Huaping, Z., Yong, L. Manipulation of disodium rhodizonate: Factors for fast-charge and fast-discharge sodium-ion batteries with long-term cyclability. *Adv. Funct. Mater.* **2016**, *26* (11), 1777.
217. Wang, C., Xu, Y., Fang, Y., Zhou, M., Liang, L., Singh, S., Zhao, H., Schober, A., Lei, Y. Extended π-conjugated system for fast-charge and -discharge sodium-ion batteries. *J. Am. Chem. Soc.* **2015**, *137* (8), 3124.
218. Wei, L., Marshall, A., Vadivukarasi, R., Xiulei, J. An organic pigment as a high-performance cathode for sodium-ion batteries. *Adv. Energy Mater.* **2014**, *4* (15), 1400554.

219. Shiwen, W., Lijiang, W., Zhiqiang, Z., Zhe, H., Qing, Z., Jun, C. All organic sodium-ion batteries with $Na_4C_8H_2O_6$. *Angewandte Chemie International Edition* **2014**, *53* (23), 5892.
220. Kim, H., Kwon, J. E., Lee, B., Hong, J., Lee, M., Park, S. Y., Kang, K. High energy organic cathode for sodium rechargeable batteries. *Chem. Mater.* **2015**, *27* (21), 7258.
221. Deng, W., Liang, X., Wu, X., Qian, J., Cao, Y., Ai, X., Feng, J., Yang, H. A low cost, all-organic Na-ion battery based on polymeric cathode and anode. *Scientific Reports* **2013**, *3*, 2671.
222. Zhao, R., Zhu, L., Cao, Y., Ai, X., Yang, H. X. An aniline-nitroaniline copolymer as a high capacity cathode for Na-ion batteries. *Electrochem. Commun.* **2012**, *21*, 36.
223. Han, S. C., Bae, E. G., Lim, H., Pyo, M. Non-crystalline oligopyrene as a cathode material with a high-voltage plateau for sodium ion batteries. *J. Power Sources* **2014**, *254*, 73.
224. Heng-guo, W., Shuang, Y., De-long, M., Xiao-lei, H., Fan-lu, M., Xin-bo, Z. Tailored aromatic carbonyl derivative polyimides for high-power and long-cycle sodium-organic batteries. *Adv. Energy Mater.* **2014**, *4* (7), 1301651.
225. Xu, F., Wang, H., Lin, J., Luo, X., Cao, S.-A., Yang, H. Poly(anthraquinonyl imide) as a high capacity organic cathode material for Na-ion batteries. *J. Mater. Chem. A* **2016**, *4* (29), 11491.
226. Zhiping, S., Yumin, Q., Tao, Z., Minoru, O., Haoshen, Z. Poly(benzoquinonyl sulfide) as a high-energy organic cathode for rechargeable Li and Na batteries. *Adv. Sci.* **2015**, *2* (9), 1500124.
227. Sakaushi, K., Hosono, E., Nickerl, G., Gemming, T., Zhou, H., Kaskel, S., Eckert, J. Aromatic porous-honeycomb electrodes for a sodium-organic energy storage device. *Nat. Commun.* **2013**, *4*, 1485.

5

Electrolytes, Additives, and Binders for Sodium Ion Batteries

5.1 Introduction

Due to the limited reserve of lithium on the earth and continuously increasing cost for the lithium-based materials, the sodium ion battery (SIBs) are attracting increasing research attention, and it is a crucial technique after the lithium ion battery(LIBs) which can be a further support to the human energy requirements in the future. The advantages of SIBs can be included as follows: the low-cost of the sodium-based electrode materials due to the rich distribution of Na, scalable fabrication technique, and promising electrochemical performance. As a result, SIB is a promising candidate as the novel energy storage system to replace the currently commercialized LIBs for a long time. More and more research groups are focusing on the commercial application of SIBs and exploring the novel promising Na-based electrodes. Unfortunately, the practical application of Na-based SIBs with outstanding electrochemical performance is also facing great challenges, including the design of the promising electrode candidates, suitable electrolytes, additives, and binders. Till now, although great research attention has been paid to developing novel electrodes for SIBs, the research efforts which are dissolving and discovering suitable electrolytes and binders are still in the deep water.[1] According to the previous research on LIBs, it is obvious that employing suitable electrolytes and binders as the necessary materials to match the relevant electrodes in the full-cell fabrication is also significantly important for the commercial battery application, because both the cathode and anode materials are immersed within the electrolyte means the interactions between the electrolyte and electrodes are crucial for the overall performance of the full batteries. For example, after the interactions between the electrolytes and electrode, the protective layer and solid electrolyte interface (SEI) films will be generated on the outer surface of the electrodes which are crucial for the cycling stability and rate performance of the full cells.[2] As a result, developing promising electrolytes and binders plays an important role to further improve the electrochemical performance of the SIBs.

In the following part, the research of the electrolytes which were used as the ingredients for the SIBs will be concluded. Moreover, the research development of the binders along with their influence on the overall battery performance also will be briefly discussed.

5.2 Electrolytes

There are some necessary requirements and key points for the SIB-based electrolytes, including the chemical/electrochemical/thermal inertness, electrical insulating, ionically conductive, low-cost, scalable production available, environmentally friendly, and facial fabrication process.[3,4] As we all know, the natural features of the liquid electrolytes have a tight relationship with the dissolved salts, the organic solvents, and also the introduced additives. Compared to the additives, the battery performance is more likely to be effected by the dissolved salts and the liquid solvents. For the suitable salts which are dissolved inside the solvent, some necessary key points are required, including the solubility within the suitable solvents, chemical stability during the oxidation and reduction conditions, and chemical inertness within the cells.[2] As for the solvent, it should be polar even under high dielectric constant, shows low viscosity to enhance the transfer of ions, electrochemically stable during the charging/discharging process, and maintains liquid status under broad temperature ranges.[2] Although tremendous research attention has been paid to developing suitable electrolytes for SIBs during the past few years, the commercialized SIB electrolytes are still not available. Compared with all the other electrolyte candidates, the carbonate ester-typed compounds which are composed of the sodium salts are regarded as the most promising ones for the practical SIBs application.

At the beginning of the research for the SIBs electrolyte, NaI, NaClO$_4$, and NaPF$_6$ are first proposed in the 1980s and employed as the salts to be dissolved in the organic solvents.[5] After that, both the NaClO$_4$ and NaPF$_6$ were considered as the most promising salts for SIB electrolytes.[3] Unfortunately, NaClO$_4$ has some disadvantages when used as the active salt for SIBs, including toxic ability, long time for drying, and safety concerns under high working voltages. However, considering of the combination historical and low-cost, the NaClO$_4$ is still the most commonly used salt in the research of SIB electrolytes.[4]

Hagenmuller's group reported the electrochemical performance of the Na$_x$CoO$_2$ electrode when NaClO$_4$ was used as the salt dissolved within the propylene carbonate electrolyte.[6] The sodium storage ability of the hard carbon was also studied by Dahn's group. They employed the ethylene carbonate: diethyl carbonate(EC:DEC) (30:70, v/v) as the liquid organic solvent and NaClO$_4$ as the salt.[7] Tirado's group studied the electrochemical ability of the NaClO$_4$/NaPF$_6$-based electrolyte with the cyano ethylene carbonate: dimethyl carbonate (CEC:DMC) (1:1) as the solvent.[8] This design endowed the SIBs with a higher capacity and increased initial Coulombic efficiency. Fujiwara's group studied the difference of the electrochemical performance of the hard carbon-based SIBs with different electrolyte solvents, such as EC, EC:DMC (1:1), butylene carbonate (BC), propylene carbonate (PC),

EC/ethylmethyl carbonate (EMC) (1:1), and PC:VC (98:2).[9] The salt used in their work was the 1 M NaClO$_4$. Compared with other kinds of electrolytes, the PC, EC:DEC, and EC showed better performances. Based on their research, the positive effect of Vinylene carbonate (VC) which was commonly accepted for the energy storage was not playing the same role. They found that with the addition of VC, there was no positive improvement for the final electrochemical performance. The ethyl methanesulfonate-typed (EMS) organic electrolyte which was dissolved with the NaClO$_4$ salt was also studied by Sun's group (Figure 5.1a and 5.1b).[10] With the introduction of the EMC to replace the conventional PC, the novel electrolyte-based SIBs showed improved anodic stability and higher ionic conductivity.

In order to improve the safety of the practical SIBs, the alternative sodium-based salts were explored to realize the commercial application of SIBs. Palacin's group systematically studied several promising electrolytes with

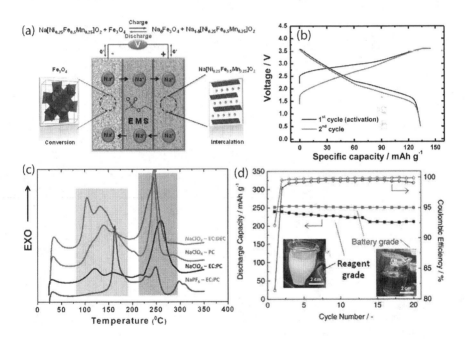

FIGURE 5.1
(a) Scheme of the C-Fe$_3$O$_4$/NaClO$_4$-ethyl methanesulfonate + 2 vol% FEC/Na[Ni$_{0.25}$Fe$_{0.5}$Mn$_{0.25}$]O$_2$ full sodium-ion battery, (b) first two charge-discharge cycles at 0.1 C-rate (13 mA g^{-1}, 0.5–3.6 V) at 25°C (Reprinted with permission from Oh, Myung, S.M. et al., *Nano Lett.*, 14, 1620–1626, 2014. Copyright 2014 American Chemical Society.), (c) DSC results of the carbonelectrode in various electrolyte formulations after the fully Na insertion process (Ponrouch, A. et al., *Energ. Environ. Sci.*, 5, 8572–8583, 2012. Reproduced by permission of The Royal Society of Chemistry.), and (d) comparison of cycling stability of hard carbon-based electrode between reagent grade and battery grade NaPF$_6$ PC solution. The inserted are the photographs of the electrolyte solution employed in this work (Reprinted with permission from Kubota, K. et al., *J. Electrochem. Soc.*, 162, A2538–A2550, 2015. Copyright 2015 The Electrochemical Society).

different salts and solvents inside to eliminate the negative interaction between the electrode and electrolytes, leading to the improved electrochemical performance and safety of the SIBs based on the optimized electrolyte formulation.[11] The viscosity, chemical/physical/electrochemical/thermal stabilities, and ionic conductivity of the different electrolytes including the diverse sodium salt-based solvents were investigated in their work. For the safety ability of the SIBs in which the hard carbon was employed as the active electrodes, they studied the thermal stability with the help of the differential scanning calorimetry (DSC) measurements. According to their study, the $NaPF_6$ within the EC/PC electrolyte showed the best thermal stability and reduced heat production during the cycling test (Figure 5.1c). The improved thermal stability of this electrolyte was mainly due to the generation of a more stable SEI film on the surface of the hard carbon electrode, leading to the improved cycling stability and increased initial Coulombic efficiency (CE). In addition, Fujiwara's group also studied the difference of the electrochemical performance between the $NaPF_6$-based and $NaClO_4$-based electrolyte.[9] Adelhelm's group conducted the relevant research of the electrolytes which were composed of $NaClO_4$-, $NaPF_6$-, and $NaCF_3SO_3$-based salts in the EC/DMC.[12] Based on their research, the $NaPF_6$-based electrolyte showed an increased ionic conductivity, and this electrolyte preferred to generate more stable SEI films on the surface of the Co-based cathode. Palacin's group also obtained the outstanding rate performance of the full symmetric SIBs with $Na_3V_2(PO_4)_2F_3$ as both the cathode and anode in the $NaPF_6$-based electrolyte with EC:PC:DMC as the solvent.[13] Unfortunately, due to the formation of the NaF during the sodiation/desodiation process, the poor ionic conductivity of this modified electrolyte appeared. Moreover, the purity of the dissolved salts within the solvents also plays a crucial role for the final performance of the cell. Komaba's group explored the electrochemical differences of the $NaPF_6$-based electrolyte with different purities.[14] For the reagent grade $NaPF_6$ salt, it was difficult to dissolve within the solvent, with part of the powder undissolved after a long time, while a better result was found for the battery grade salts, leading to transparent liquid after the $NaPF_6$ dissolution. Due to the better dissolution of the battery grade $NaPF_6$, it showed better electrochemical performance than that of the reagent grade salt (Figure 5.1d). Therefore, the purity of the salts and the proportion of the ingredients inside the electrolyte are critical for the battery performance.

Just during the past few years, the ether-typed electrolytes are also attracting great research attention and an increasing number of studies have been reported about the SIBs using ether-based electrolytes.[15–18] For example, the co-insertion of the Na ions and organic solvent can be obtained when using the ether-based electrolytes for the SIBs. Chen and Kang systematically investigated the electrochemical performance of graphite anode-based SIBs using different ether-typed electrolytes. By adjusting the compositions within the ether-based electrolytes and also the working voltage ranges, the transition-metal disulfides can deliver better rate capability and improved cycling stability. For these transition-metal chalcogenides, it is promising to employ the ether-based electrolytes because they showed enhanced chemical/electrochemical stability and higher reaction activity than using the carbonate-based electrolytes.[15]

5.3 Additives

Apart from the salts and solvents, another important component inside the electrolyte is the additive. With the introduction of additives, it provides us a novel way to produce the functional electrolytes with some unique features. For example, improved stable SEI films between the electrolyte and electrode, enhanced safety of the full cells, and electrochemical stability of the electrolytes can be obtained via using extra additives.[19,20]

The comparison of the electrochemical performance among different additives, such as the fluoroethylene carbonate (FEC), 4,5-difluoroethylene carbonate (DFEC), ethylene sulfite (ES), and VC, within the NaPF$_6$-based PC electrolyte was studied by Ohsawa's group.[20] All of these additives have been successfully used as the additives in the LIBs. According to this work, it was found that only the FEC (2%) exhibited positive effects when used as additive for both the anode and cathode. This improvement can be explained by the stable formation of the SEI films on the electrodes' surface and the suppression of the negative reaction between electrolyte and electrode (Figure 5.2a and 5.2b). In addition, the electrolyte maintained a better chemical stability with the introduction of FEC under high charging voltage, leading to

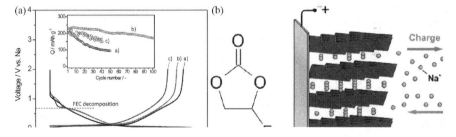

FIGURE 5.2
(a) The first oxidation/reduction profiles for hard carbon-based electrodes in 1 M NaClO$_4$ PC electrolyte with or without the addition of FEC. Inset exhibits variation in reversible oxidative capacities for the hard carbon during successive cycle test, and (b) the schematic representation of the cell. (Reprinted with permission from Komaba, S. et al., *ACS Appl. Mater. Inter.*, 3, 4165–4168, 2011. Copyright 2011 American Chemical Society.)

the good cycling stability of SIBs. Apart from the PC-only-typed electrolyte, much improved performance of SIBs can be further obtained in the binary system which was composed of the EC and PC.[21] With the help of EC, stable and efficient SEI films were generated on the hard carbon surface without the addition of FEC. Komaba's group studied the sodium storage ability of hard carbon in the $NaClO_4$ and $NaPF_6$ dissolved PC/EC electrolyte with or without FEC additives.[22] A new discharge potential platform appeared at about 0.7 V during the initial cycling test when the FEC was used as additive in both the $NaClO_4$ and $NaPF_6$-based electrolytes. The newly formed voltage plateau was mainly due to the decomposition of the introduced FEC. Therefore, a higher reversible capacity was also obtained when compared with the SIBs without FEC additive.

Moreover, the effect from the electrolyte additives can also be found for the alloying/conversion-based reaction anodes.[22] For these kinds of compounds, large volume changes usually occurred during the charging/discharging process, therefore, a stable SEI film plays a more important role for the overall cycling stability of these kinds of anodes. Due to the formation of a more stable SEI film, the structural stability of the electrode and the transfer of Na ions within the electrode are both improved.

As a result, the FEC-added electrolyte shows promising performance, and the research about the difference of the composition for the FEC-triggered SEI films was also becoming another hot topic. According to the XPS test, a high concentration of compounds, such as the inorganic salts, C=O rich materials, and alkyl carbonates which are unstable during the cycling test, could be found inside the SEI films without the FEC additive.[23] Komaba's group investigated the possible electrochemical reaction mechanism of the FEC reduction inside the SIBs during the cycling.[24] In addition, Shenoy's group reported the reaction process of the FEC-added electrolyte during the reduction stage.[25] They believed that FEC would be reduced at the beginning of the reaction and then a protective SEI film which was derived from polymerized organic function groups would be generated.[26-28] Kobama's group also studied the interface between the electrode and electrolyte and found that positive influence could be obtained when the VC was added as an additive in the electrolyte.[24] Due to the introduction of VC, the reversible capacity and cycling stability of the black P anode were both significantly improved.[20]

Therefore, it can be summarized that the most promising electrolyte for the commercial available SIBs should include the following points: (1) $NaPF_6$ dissolved carbonate-ester binary or ternary solvents are the most promising electrolytes for SIBs and have been commonly used in most of the research about SIBs. (2) For the organic solvents, the PC is crucial and necessary in both the binary and ternary systems because of its high dielectric constant and wide operating electrochemical and operation temperature ranges. (3) With the introduction of the FEC additive, stable SEI films were generated on the surface of electrodes, which can significantly improve the structural stability of the electrodes during the charging/discharging process. (4) In order to achieve outstanding electrochemical performance of the SIBs, we need to employ the battery grade solvents and salts with the highest purity.

5.4 Binders

Binders are an important part for assembling batteries, especially for the anode electrode fabrication. Due to the large volume changes of most anode materials, a suitable binder could effectively address this problem via the strong chemical bonding between binder and electrode, leading to the improved structural stability. In addition, with the introduction of the binder, the active electrode materials can have a tight contact with the current collector, avoiding the stripping during the cycling process. Moreover, appropriate binders are important to obtain a stable SEI film on the outer surface of the electrodes.

According to the previous research and also the currently commercialized LIBs, poly(vinylidene fluoride) (PVDF) is the most important binder in the energy storage systems, due to its strong chemical bonding and stable chemical status under wide surroundings.[29]

Unfortunately, the cost for PVD during the commercial production is very high, and there are some toxic organic solvents that are necessary for PVDF production, therefore, exploring novel binders with green and facial fabrication methods, low-cost, and outstanding performance when used as binders for batteries is becoming important.[30–32]

With the above concerns, the green and water soluble organic binder attracted great research interest, such as the poly(acrylic acid) (PAA), sodium carboxymethyl cellulose (Na-CMC), and sodium alginate (Na-Alg) (Figure 5.3a and b).[32–41] These kinds of materials are preferred to be used as the binder for alloying-based anode materials to improve their structural stability and the following cycling stability, because they have a large volume expansion during the Na/Li ions insertion process. With the help of these binders, a thermally interconnected 3D framework was generated which can accommodate the volume changes of the active electrode materials.

Derived from the purification of cellulose, Na-CMC is a promising candidate because it has green and low-cost advantages. Especially, Na-CMC is helpful for the formation of the stable SEI film on the outer surface of electrodes, leading to the high initial CE and long cycling life.[42] Na-Alg is another organic material from brown algae with high-modulus, and it provides stronger

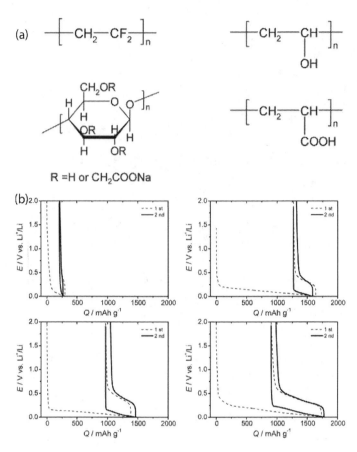

FIGURE 5.3
(a) Structure of four different kinds of polymer binders: PVDF, PVA, CMC-NA, PAA, and (b) chronopotentiograms of the SiO powder electrodes prepared with PVdF, PVA, CMCNa, and PAA binders in 1 mol dm^{-3} LiPF$_6$ EC/DMC solution at a rate of 100 mA g^{-1}. (Reprinted with permission from Komaba, S. et al., *J. Phys. Chem. C*, 115, 3487–13495, 2011. Copyright 2011 American Chemical Society.)

interfacial reaction between the active electrode particles and organic binder molecules due to its more polar ability.[34] PAA is also a promising binder, because it is endowed with the carboxylic groups in the molecules, which can provide stronger interaction and binding strength.[41] In addition, a more uniform and stable SEI film can be produced on the surface of an electrode when the PAA is used as a binder during the electrode fabrication.[40] Moreover, compared with the PVDF binder, the decomposition of the electrolyte was suppressed for the PAA binder.[43] Komaba's group compared the electrochemical performance of the hard carbon electrode-based SIBs using the

Electrolytes, Additives, and Binders for Sodium Ion Batteries 139

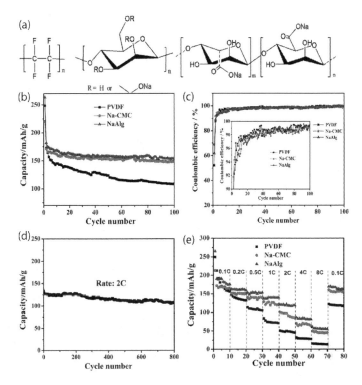

FIGURE 5.4
(a) The chemical structure of different binders. Electrochemical performance of $Na_{0.11}Li_{3.89}Ti_5O_{12}$ electrodes with different binders, (b) cyclic performance of PVDF, Na-CMC, and NaAlg electrodes, rate: 0.1C (17.5 mA g^{-1}), (c) Coulombic efficiency versus cycle number plots of PVDF, Na-CMC, and NaAlg electrodes, (d) cycling stability of NaAlg electrodes, rate: 2C (350 mA g^{-1}), and (e) rate performance of poly vinylidene fluoride(PVDF), Na-sodium carboxymethyl cellulose, and NaAlg electrodes. (Reprinted with permission from Zhao, F. et al., *J. Electrochem. Soc.*, 163, A690–A695, 2016. Copyright 2011 Electrochemical Society).

CMC and PVDF binders. They found that the decomposition of the electrolyte was effectively reduced with the help of the CMC binder due to the uniform SEI films generated on the surface of the electrode.[44] In order to improve the structural stability and cycling life, Huang's group employed CMC as the binder for the $Li_4Ti_5O_{12}$-based SIBs. Compared with the PVDF binder, the CMC-based cells showed higher initial CE and increased capacity retention over 100 cycling test (Figure 5.4a–5.4e).[45] The PAA binder was also used by Lee's group for the alloying-based anode in the SIBs. The PAA binder exhibited outstanding ability to address the large volume changes of the alloying-based anodes during the sodiation/desodiation stages, resulting in the long cycling life.

References

1. Vignarooban, K., Kushagra, R., Elango, A., Badami, P., Mellander, B. E., Xu, X., Tucker, T. G., Nam, C., Kannan, A. M., Current trends and future challenges of electrolytes for sodium-ion batteries. *Int. J. Hydrogen Energ.* **2016**, *41* (4), 2829–2846.
2. Ponrouch, A., Monti, D., Boschin, A., Steen, B., Johansson, P., Palacin, M. R., Non-aqueous electrolytes for sodium-ion batteries. *J. Mater. Chem. A* **2015**, *3* (1), 22–42.
3. Xu, K., Electrolytes and interphases in Li-ion batteries and beyond. *Chem. Rev.* **2014**, *114* (23), 11503–11618.
4. Aurbach, D., Talyosef, Y., Markovsky, B., Markevich, E., Zinigrad, E., Asraf, L., Gnanaraj, J. S., Kim, H. J., Design of electrolyte solutions for Li and Li-ion batteries: A review. *Electrochim. Acta* **2004**, *50* (2–3), 247–254.
5. Newman, G. H., Klemann, L. P., Ambient-temperature cycling of an Na-Tis$_2$ cell. *J. Electrochem. Soc.* **1980**, *127* (10), 2097–2099.
6. Braconnier, J. J., Delmas, C., Fouassier, C., Hagenmuller, P., Electrochemical behavior of the phases NaxCoO2. *Mater. Res. Bull.* **1980**, *15* (12), 1797–1804.
7. Stevens, D. A., Dahn, J. R., High capacity anode materials for rechargeable sodium-ion batteries. *J. Electrochem. Soc.* **2000**, *147* (4), 1271–1273.
8. Alcantara, R., Lavela, P., Ortiz, G. F., Tirado, J. L., Carbon microspheres obtained from resorcinol-formaldehyde as high-capacity electrodes for sodium-ion batteries. *Electrochem. Solid. St.* **2005**, *8* (4), A222–A225.
9. Komaba, S., Murata, W., Ishikawa, T., Yabuuchi, N., Ozeki, T., Nakayama, T., Ogata, A., Gotoh, K., Fujiwara, K., Electrochemical Na insertion and solid electrolyte interphase for hard-carbon electrodes and application to Na-Ion batteries. *Adv. Funct. Mater.* **2011**, *21* (20), 3859–3867.
10. Oh, S. M., Myung, S. T., Yoon, C. S., Lu, J., Hassoun, J., Scrosati, B., Amine, K., Sun, Y. K., Advanced Na[Ni$_{0.25}$Fe$_{0.5}$Mn$_{0.25}$]O-2/C-Fe$_3$O$_4$ sodium-ion batteries using EMS electrolyte for energy Storage. *Nano Lett.* **2014**, *14* (3), 1620–1626.
11. Ponrouch, A., Marchante, E., Courty, M., Tarascon, J. M., Palacin, M. R., In search of an optimized electrolyte for Na-ion batteries. *Energ. Environ. Sci.* **2012**, *5* (9), 8572–8583.
12. Bhide, A., Hofmann, J., Durr, A. K., Janek, J., Adelhelm, P., Electrochemical stability of non-aqueous electrolytes for sodium-ion batteries and their compatibility with Na$_{0.7}$CoO$_2$. *Phys. Chem. Chem. Phys.* **2014**, *16* (5), 1987–1998.
13. Ponrouch, A., Dedryvere, R., Monti, D., Demet, A. E., Mba, J. M. A., Croguennec, L., Masquelier, C., Johansson, P., Palacin, M. R., Towards high energy density sodium ion batteries through electrolyte optimization. *Energ. Environ. Sci.* **2013**, *6* (8), 2361–2369.
14. Kubota, K., Komaba, S., Review-practical issues and future perspective for Na-ion batteries. *J. Electrochem. Soc.* **2015**, *162* (14), A2538–A2550.
15. Kajita, T., Itoh, T., Ether-based solvents significantly improved electrochemical performance for Na-ion batteries with amorphous GeO$_x$ anodes. *Phys. Chem. Chem. Phys.* **2017**, *19* (2), 1003–1009.
16. Zhang, K., Hu, Z., Liu, X., Tao, Z. L., Chen, J., FeSe$_2$ microspheres as a high-performance anode material for Na-ion batteries. *Adv. Mater.* **2015**, *27* (21), 3305–3309.

17. Zhang, K., Park, M. H., Zhou, L. M., Lee, G. H., Li, W. J., Kang, Y. M., Chen, J., Urchin-like CoSe$_2$ as a high-performance anode material for sodium-ion batteries. *Adv. Funct. Mater.* **2016**, *26* (37), 6728–6735.
18. Zhang, K., Park, M., Zhou, L. M., Lee, G. H., Shin, J., Hu, Z., Chou, S. L., Chen, J., Kang, Y. M., Cobalt-doped FeS$_2$ nanospheres with complete solid solubility as a high-performance anode material for sodium-ion batteries. *Angew. Chem. Int. Edit.* **2016**, *55* (41), 12822–12826.
19. Kim, H., Hong, J., Park, K. Y., Kim, H., Kim, S. W., Kang, K., Aqueous rechargeable Li and Na ion batteries. *Chem. Rev.* **2014**, *114* (23), 11788–11827.
20. Komaba, S., Ishikawa, T., Yabuuchi, N., Murata, W., Ito, A., Ohsawa, Y., Fluorinated ethylene carbonate as electrolyte additive for rechargeable Na batteries. *ACS Appl. Mater. Inter.* **2011**, *3* (11), 4165–4168.
21. Irisarri, E., Ponrouch, A., Palacin, M. R., Review-hard carbon negative electrode materials for sodium-ion batteries. *J. Electrochem. Soc.* **2015**, *162* (14), A2476–A2482.
22. Dahbi, M., Nakano, T., Yabuuchi, N., Fujimura, S., Chihara, K., Kubota, K., Son, J. Y., Cui, Y. T., Oji, H., Komaba, S., Effect of hexafluorophosphate and fluoroethylene carbonate on electrochemical performance and the surface layer of hard carbon for sodium-ion batteries. *Chemelectrochem* **2016**, *3* (11), 1856–1867.
23. Baggetto, L., Ganesh, P., Meisner, R. P., Unocic, R. R., Jumas, J. C., Bridges, C. A., Veith, G. M., Characterization of sodium ion electrochemical reaction with tin anodes: Experiment and theory. *J. Power Sources* **2013**, *234*, 48–59.
24. Dahbi, M., Yabuuchi, N., Fukunishi, M., Kubota, K., Chihara, K., Tokiwa, K., Yu, X. F., Ushiyama, H., Yamashita, K., Son, J. Y., Cui, Y. T., Oji, H., Komaba, S., Black phosphorus as a high-capacity, high-capability negative electrode for sodium-ion batteries: Investigation of the electrode/interface. *Chem. Mater.* **2016**, *28* (6), 1625–1635.
25. Kumar, H., Detsi, E., Abraham, D. P., Shenoy, V. B., Fundamental mechanisms of solvent decomposition involved in solid-electrolyte interphase formation in sodium ion batteries. *Chem. Mater.* **2016**, *28* (24), 8930–8941.
26. Chen, X. L., Li, X. L., Mei, D. H., Feng, J., Hu, M. Y., Hu, J. Z., Engelhard, M., Zheng, J. M., Xu, W., Xiao, J., Liu, J., Zhang, J. G., Reduction mechanism of fluoroethylene carbonate for stable solid-electrolyte interphase film on silicon anode. *Chemsuschem* **2014**, *7* (2), 549–554.
27. Markevich, E., Fridman, K., Sharabi, R., Elazari, R., Salitra, G., Gottlieb, H. E., Gershinsky, G., Garsuch, A., Semrau, G., Schmidt, M. A., Aurbach, D., Amorphous columnar silicon anodes for advanced high voltage lithium ion full cells: Dominant factors governing cycling performance. *J. Electrochem. Soc.* **2013**, *160* (10), A1824–A1833.
28. Etacheri, V., Haik, O., Goffer, Y., Roberts, G. A., Stefan, I. C., Fasching, R., Aurbach, D., Effect of fluoroethylene carbonate (FEC) on the performance and surface chemistry of Si-Nanowire Li-ion battery anodes. *Langmuir* **2012**, *28* (1), 965–976.
29. Yabuuchi, N., Kubota, K., Dahbi, M., Komaba, S., Research development on sodium-ion batteries. *Chem. Rev.* **2014**, *114* (23), 11636–11682.
30. Cai, Z. P., Liang, Y., Li, W. S., Xing, L. D., Liao, Y. H., Preparation and performances of LiFePO$_4$ cathode in aqueous solvent with polyacrylic acid as a binder. *J. Power Sources* **2009**, *189* (1), 547–551.

31. Lux, S. F., Schappacher, F., Balducci, A., Passerini, S., Winter, M., Low cost, environmentally benign binders for Lithium-ion batteries. *J. Electrochem. Soc.* **2010**, *157* (3), A320–A325.
32. Komaba, S., Shimomura, K., Yabuuchi, N., Ozeki, T., Yui, H., Konno, K., Study on polymer binders for high-capacity SiO negative electrode of Li-ion batteries. *J. Phys. Chem. C* **2011**, *115* (27), 13487–13495.
33. Ming, J., Ming, H., Kwak, W. J., Shin, C., Zheng, J. W., Sun, Y. K., The binder effect on an oxide-based anode in lithium and sodium-ion battery applications: The fastest way to ultrahigh performance. *Chem. Commun.* **2014**, *50* (87), 13307–13310.
34. Kovalenko, I., Zdyrko, B., Magasinski, A., Hertzberg, B., Milicev, Z., Burtovyy, R., Luzinov, I., Yushin, G., A major constituent of brown algae for use in high-capacity Li-ion batteries. *Science* **2011**, *334* (6052), 75–79.
35. Li, Y. M., Hu, Y. S., Titirici, M. M., Chen, L. Q., Huang, X. J., Hard carbon microtubes made from renewable cotton as high-performance anode material for sodium-ion batteries. *Adv. Energy Mater.* **2016**, *6* (18).
36. Palomares, V., Casas-Cabanas, M., Castillo-Martinez, E., Han, M. H., Rojo, T., Update on Na-based battery materials: A growing research path. *Energ. Environ. Sci.* **2013**, *6* (8), 2312–2337.
37. Koo, B., Kim, H., Cho, Y., Lee, K. T., Choi, N. S., Cho, J., A highly cross-linked polymeric binder for high-performance silicon negative electrodes in lithium ion batteries. *Angew. Chem. Int. Edit.* **2012**, *51* (35), 8762–8767.
38. Zhao, X., Vail, S. A., Lu, Y. H., Song, J., Pan, W., Evans, D. R., Lee, J. J., Antimony/graphitic carbon composite anode for high-performance sodium-ion batteries. *ACS Appl. Mater. Inter.* **2016**, *8* (22), 13871–13878.
39. Ui, K., Kikuchi, S., Mikami, F., Kadoma, Y., Kumagai, N., Improvement of electrochemical characteristics of natural graphite negative electrode coated with polyacrylic acid in pure propylene carbonate electrolyte. *J. Power Sources* **2007**, *173* (1), 518–521.
40. Magasinski, A., Zdyrko, B., Kovalenko, I., Hertzberg, B., Burtovyy, R., Huebner, C. F., Fuller, T. F., Luzinov, I., Yushin, G., Toward efficient binders for Li-ion battery Si-based anodes: Polyacrylic acid. *ACS Appl. Mater. Inter.* **2010**, *2* (11), 3004–3010.
41. Chou, S. L., Pan, Y. D., Wang, J. Z., Liu, H. K., Dou, S. X., Small things make a big difference: Binder effects on the performance of Li and Na batteries. *Phys. Chem. Chem. Phys.* **2014**, *16* (38), 20347–20359.
42. Dahbi, M., Nakano, T., Yabuuchi, N., Ishikawa, T., Kubota, K., Fukunishi, M., Shibahara, S., Son, J. Y., Cui, Y. T., Oji, H., Komaba, S., Sodium carboxymethyl cellulose as a potential binder for hard-carbon negative electrodes in sodium-ion batteries. *Electrochem. Commun.* **2014**, *44*, 66–69.
43. Mazouzi, D., Karkar, Z., Hernandez, C. R., Manero, P. J., Guyomard, D., Roue, L., Lestriez, B., Critical roles of binders and formulation at multiscales of silicon-based composite electrodes. *J. Power Sources* **2015**, *280*, 533–549.
44. Zhang, W. J., Dahbi, M., Komaba, S., Polymer binder: A key component in negative electrodes for high-energy Na-ion batteries. *Curr. Opin. Chem. Eng.* **2016**, *13*, 36–44.
45. Zhao, F., Xue, P., Ge, H. H., Li, L., Wang, B. F., Na-doped $Li_4Ti_5O_{12}$ as an anode material for sodium-ion battery with superior rate and cycling performance. *J. Electrochem. Soc.* **2016**, *163* (5), A690–A695.

6
Current Challenges and Future Perspectives

6.1 Introduction

In recent years, sodium ion batteries (SIBs) have been identified as a potential candidate for grid electricity storage application even though they are still at the developing phase. Also, the SIBs can benefit from the development made in the technologies of well-established LIBs. Recent advancement in the field of SIBs suggests that these systems could utilize aqueous electrolytes, this could bring down the cost effectively.[1,2] Furthermore, sodium is the second lightest and smallest alkali metal, the first being lithium, thus the energy density sacrifice would be minimal as compared to other developing secondary batteries, such as, magnesium ion batteries, potassium ion batteries, etc.[3] Moreover, unlike lithium, sodium metal does not form an alloy with aluminum so the expensive copper on the anode side can be replaced with the low-cost aluminum foil. Another advantage is that, unlike LIB, the risk associated with the transportation of SIB is minimal. With all the aforementioned cost reduction techniques that can be employed in SIBs, the atomic weight and the standard electrode potential of sodium as compared to lithium is always a bottleneck. This bulky size and the lower potential could adversely affect the stability and energy of the SIBs. Thus, it is necessary to understand the underlying problems, and at the same time the potential advantages of the SIBs, to bring this potential system to the market. Even though the SIBs were invented in the early 1980s in parallel to LIBs, there are very limited studies on the full-cell aspect; however, there are thousands of reports on the promising electrochemical properties of SIBs in half-cell format. For instance, even in the 1980s a few United States and Japanese companies developed full-cell, where P2-type Na_xCoO_2 was used as the cathode and sodium-lead alloy as the anode.[4,5] Also, these studies were reported even before the commercialization of LIBs. This full-cell format exhibited a promising cycle life of 300 cycles, however, it could only show a low average discharge potential of <3.0 V, thus it could not match the performance of the front-line LIBs (carbon/$LiCoO_2$ cell) in that period. Nevertheless, in recent years the SIB technologies have been renewed due to various reasons, including the limited availability of lithium resources and its increasing cost. As discussed in the previous chapters, there are plenty of positive (transition metal oxides, transition metal fluorides, phosphates, sulfates, Prussian blue analogues, organic materials, etc.)[6–9] and negative electrodes (hard carbon, alloys, phosphorous materials, phosphides, organic carboxylates, etc.)[10–17] that have been investigated and have shown potential half-cell performances. Although graphite and silicon anodes are proven to be a potential candidate for LIBs, but these negative electrodes are practically electrochemically inactive in SIBs. This is applicable for many other cathode systems explored in SIBs as well. Thus, it is not advisable to take all the proven technologies of LIBs in SIBs blindly. Hence, while dealing with materials for SIBs, there should be always some careful understanding and analysis. Moreover, in spite of the advancement in cathode and anode materials, there remain numerous challenges, including cell design, understanding of fundamental electrochemical mechanism, poor initial Coulombic efficiency, electrode mass balancing, functional full-cell fabrication, etc., in the field of SIBs, that are obstructing the practical application. This chapter discusses the current challenges in various fields and recommends some future directions for SIBs; this could provide significant insights into fundamental and practical issues in the progress of SIBs.

6.2 Current Challenges

There are plenty of challenges when it comes to the practical usage of SIBs. For instance, the preparation of SIB half-cell (to analyze the individual electrode performance) should be much more careful as compared to the LIB counterpart because of the high reactivity of the sodium metal as compared to the lithium. Thus, the working conditions should be carefully maintained throughout the cell assembly, such as proper water and oxygen levels in the glove box, purity of the argon gas, etc. Also, the formation of the stable passivation solid electrolyte interfaces (SEI) layer on sodium metal (in half-cell format) or negative electrode materials (in full-cell format) is very difficult.

6.2.1 Electrode Selection

Electrode materials with an open crystal framework are more desirable in SIBs, because the larger diameter of the sodium (as compared to the lithium) should be lodged in the crystal framework. The main concern with the electrode material is the maintenance of the structural stability during the prolonged charge/discharge process. For instance, during the Na^+ insertion process, the original crystal structure of the material is bound to change, and trying to bring back its original state can cause disruption of the structural integrity. Apart from that, most of the sodiated transition metal compounds are hygroscopic in nature. The hydration of such material can cause a reduction in electrochemical performances due to the formation of inactive NaOH. Thus, while handling such electrode materials, immense care should be taken to avoid any kind of moisture exposure. A study conducted by Sumitomo Chemical Co., Ltd. showed that they have successfully prepared moisture free O3-type Ca-doped $NaFe_{0.4}Ni_{0.3}Mn_{0.3}O_2$ cathodes and also managed to minimize the tendency to absorb moisture by the Ca doping approach. It is generally believed that due to the larger size of the Na^+ ions, the diffusion kinetics in SIBs could be difficult as compared to LIBs. However, an experimental and computational study conducted by Ceder's group demonstrated that the Na^+ extraction/insertion from/into sodium-based materials is relatively faster as compared to the Li-containing materials due to the low Lewis acidity of Na^+ as compared to the Li^+. The studies have shown that the O3-type $Na_{1-x}FeO_2$ compounds having the Fe^{3+}/Fe^{4+} redox couple is electrochemically active (unlike the lithium counterpart). Also, the irreversible capacity loss due to the iron migration can be avoided by the substitution of Fe in the transition metal site with Co or Mn.[18–20] Even with all these advantages, the insufficient cycling performance has yet to be solved. For the polyanionic materials, the superior thermal stability (due to the presence of P–O covalent bonds in the crystal structure) as compared to the oxide material, is always a positive attribute. The Sodium super ion conductor (NASICON)-type compounds have high Na^+ ionic conductivity because of their 3D open crystal framework. Many studies on NASICON-type $Na_3V_2(PO_4)_3$ have shown superior electrochemical performances in half-cell format, however, the full-cell performance of this system is still not satisfactory. Another widely studied material for SIBs is the Prussian blue and its analogues. These materials have attracted considerable attention due to their superior electrochemical properties, such as high energy and power density. However, these materials showed poor thermal stability and a higher level of water molecules in the crystal structure (which lead to poor electrochemical performances). To minimize the presence of water molecules in the crystal structure, a controlled synthesis should be performed. Another major challenge in the SIB technology is that the Na^+ cannot be inserted to the graphite anodes (which is a proven electrode for LIBs). Thus, an alternate anode material for large-scale energy storage systems needs to be identified. Sodium metal as an anode for SIBs is not recommended because of its high reactivity, due to the formation of dendrite and hence the related safety concerns. Many other non-graphitic carbon materials such as carbon black, pitch-based carbon-fibers have been identified for SIBs. However, as of now, hard carbon is the best anode material available for SIBs.[21–23] But, the poor initial Coulombic efficiency and large irreversible capacity (during the initial cycles) are some of the challenges to be solved for this material prior to the commercialization. Also, lower operational potential of carbon-based anodes raises severe safety concerns for practical applications. Thus, titanium-based transition metal oxides are believed to be a safer anode for SIBs.[24,25] However, the research in this field is in its infancy stage. Hence, intensive research needs to be done in this area to meet the requirements for the practical cell fabrication.

6.2.2 Electrolyte, Additives, and Binder Selection

Looking back to the development of commercial LIBs, it is evident that an appropriate choice of electrolyte, additives, and binders is as important as the selection of cathodes and anodes to fabricate a fully functional battery. This is mainly due to the influence of electrolyte and binders are inevitable, while forming a protective layer, such as SEI and surface layers, at the electrode surface. Hence, identifying suitable and optimal formulation of electrolyte and binders is crucial for the development of high performing SIBs. For this, strategies and ideas should be taken from the well-established LIB technologies. There are a wide variety of electrolytes available for SIBs, such as organic, ionic, and aqueous liquid electrolytes, solid- and gel-polymer electrolytes, and inorganic solid electrolytes, and they are still at the developing phase. Among these electrolytes, the aqueous electrolyte batteries are considered to be a cost-effective solution for the energy storage system and were successfully commercialized. One of the major drawbacks of the aqueous electrolyte is its low operating potential window, leading to low energy performances. While the organic liquid electrolyte, which is based on carbonate-ester polar solvents and sodium salts, are the widely studied electrolyte for SIBs because of their superior properties, such as large voltage window, high ionic conductivity, and promising temperature performance. The main challenge with this electrolyte is its sensitivity to the water content. Thus, preparation and handling of these chemicals under a normal atmosphere is not recommended. In most cases, the normal air-exposed organic electrolyte can cause deterioration in electrochemical performances and sometimes even damage to the battery, leading to serious safety concerns. Thus, the organic electrolyte should only be handled in the Ar-filled glove box or such inert atmosphere. Another major drawback of the organic electrolyte is the chance of leading to battery explosions as a result of thermal runaway or oxygen evolution. To avoid this, the battery should be charged, stored, and transported according to the safety manual. Also, according to the company Tesla, it is better to use some advanced battery designs and technologies to reduce the risk associated with the organic electrolyte-based batteries. For instance, employ a battery design that could provide a probable pathway through a portion of the cell for the effective release of the thermal energy that formed because of thermal runaway; thus, minimizing the chances of a break in an undesirable location. Another concern for the electrolyte is anodic stability at high voltage without electrolyte decomposition, gas evolution, and battery swelling. To avoid this, a well-balanced choice of electrolyte and electrode needs to be identified. Also with the suitable electrolyte additives, a safe potential window of a battery system can be extended. Generally, some additives are used in electrolytes to form a stable SEI layer and thus to improve its electrochemical stability. Which is also used to minimize some battery issues like flammability and overcharging. Among different explored electrolytes, fluoroethylene carbonate (FEC) is the best and efficient electrolyte additive for both the electrodes. An appropriate amount of FEC additive, generally 2 volume%, is useful to form a high-quality passivation layer on the electrode surface and for reducing the side reactions between the electrode and the electrolyte. Some studies have reported that the presence of the FEC additive could lead to the decrease in specific capacity and Coulombic efficiency of the hard carbon electrode.[26,27] Thus, proper care should be employed while selecting the additive and its concentration.

Another key component to improve the electrode performance is the appropriate choice of a binder. Particularly, in anode systems, Na storage properties are dramatically influenced by the usage of appropriate binders. Thus, developing suitable binders is also an important strategy to maintain stability in the electrode surface and reduce electrode distortion during the sodium intercalation/deintercalation process. Among different binders, the widely studied and commercially available is the poly (vinylidene fluoride) (PVDF), because of its good chemical and electrochemical stabilities. But the major drawbacks of this binder are high production cost (while making slurry) and usage of toxic organic material as the solvent (N-methyl pyrrolidone). Thus, to nullify these drawbacks many water soluble binders have been explored, such as carboxymethyl cellulose (CMC), poly (acrylic acid) (PAA), and sodium alginate (Na-Alg).

6.2.3 Full-Cell Performance

Current research is concentrated mostly on sodium-ion half-cells (employing Na metal as the counter electrode). Therefore, development of practical sodium ion full-cells (without Na metal) remains a critical challenge. Recently, in order to realize the practicality of SIBs, many research groups have started focusing on the development of rational full-cell design. Furthermore, battery companies such as USA-based Aquion Energy, UK-based Faradion, and Japan-based Sumitomo Electric Industries Ltd. are constantly trying to develop functional full-cell designs. Historically, research on sodium-ion full-cell was conducted earlier than the commercialization of LIBs.[28] For instance, decades back in 1988, sodium-ion full-cells were explored based on sodium-lead alloy as the anode and the P2-type layered Na_xCoO_2 as the cathode, which showed a cycling stability of over 300 cycles. However, compared to LIBs, the sodium ion full-cells did not attract much attention because of their low average discharge potential of <3 V.[28] Nonetheless, the introduction of hard carbon anodes has caused a dramatic improvement of sodium ion full-cell. The P2-type cathode materials with a common chemical formula of Na_xMO_2 (where M = transition metal) showed promising electrochemical performances, however, when $x < 0.7$ resulted in lower initial sodium content in the crystal structure, this led to an abnormal Coulombic efficiency of above 100% in the 1st cycle.[29] Therefore, the intrinsic properties affect the practical full-cell fabrication. Many studies have reported excellent sodium-ion full-cell performances, however, most of the studies have utilized the presodiated electrodes in order to reduce the irreversible capacity loss during the initial cycles.[30,31] Also, these presodiated systems require a different electrode mass balancing to prevent sodium plating. Thus, this method is not practical for commercial applications, but just for scientific investigations, such as to measure the maximum electrochemical properties of materials in a full-cell format. In addition, such presodiated hard carbons suffer from low rate performances, and the voltage plateau would be very close to the sodium plating voltage, causing safety concerns. Thus, to obtain functional sodium ion full-cells with better safety, good rate performances, and superior cycling stability, further in-depth investigations of full-cell designs would be required, which include proper electrode mass balancing between the electrodes, appropriate selection of working potential range, and stable electrolytes. In addition, to achieve better safety without compromising the electrochemical performances, proper research on functional additives and binders are necessary, this is to ensure a uniform SEI layer formation at the electrode surfaces. Further, the total production cost of electrode active materials and the cost of other battery materials/components should be considered. Thus, to design a practical and efficient SIB, a good balance between battery performance, safety, and cost is necessary.

6.3 Future Perspectives

In the future, the demand for electric vehicles (EVs) and other energy storage applications are anticipated to increase exponentially from the year of 2020, and the rechargeable battery market will be expected to expand almost twice of the current level (i.e., about 112 billion US dollars). The expandability of LIBs is one of the options, however, the limited availability of lithium resources and the geological issue (where the lithium resources are localized in mainly South America) are the serious concerns. Use of cobalt-based or other rare materials as cathodes is another concern on the cost of batteries for future energy demand. Thus, more of the present cathode materials are expected to be consumed in the future. Here comes the necessity of the investigation of alternative secondary batteries that can substitute for the present LIBs. In fact, in the 1980s SIBs were developed along with the LIBs. Nevertheless, the better electrochemical properties of LIBs and the subsequent commercialization of LIBs by Sony in the 1990s restricted the further growth of SIBs. Whereas, the recent high demand for energy storage systems using LIBs, both in portable and EV applications, has lead the investigation of the SIB technology again. Thus, starting from 2010, research trends have shifted more toward SIBs and the investigations for new electrode systems for SIBs have been tremendously increased. In particular, the technological and scientific knowledge of LIBs are leading a fast development in the SIBs. For instance, the O3-type layered oxide materials are very attractive because of their similar crystal structure as commercial $LiCoO_2$ and Ni-rich derivatives of LIBs. However, in general, their capacities are limited to below 130 mA h g^{-1} because of the simultaneous structural changes induced due to the presence of large Na^+ in the crystal structure. On the other hand, the P2-type layered compounds have shown high capacities of about 200 mA h g^{-1}. However, this P2-type material has a serious disadvantage, that is the low extractable sodium content in the compounds (e.g., $Na_{0.7}MO_2$; M: transition metals), leading to irreversible capacity loss and abnormal Coulombic efficiency during the initial cycles. Studies have demonstrated that the appropriate sacrificing material can undergo oxidative decomposition at the first charge, to avoid the irreversible capacity at the initial cycle. This could ensure a high Coulombic efficiency of about 100%. With this advancement, the commercialization of the SIBs in the future can be expected. A scientific work utilized Na_3N and Na_3P as the sacrificing material to minimize the first cycle abnormal Coulombic efficiency of over 130% and to compensate the sodium deficiency. This work could minimize the irreversible capacity loss of P2-type cathode materials in the initial cycle, however, the first cycle Coulombic efficiency was not appropriate to fabricate the full-cell SIB as compared to the O3-type materials. Therefore, it is necessary to revisit the research on O3-type layered oxide cathode to fabricate a fully functional SIBs. Polyanionic compounds are considered to be attractive as a cathode for SIBs, due to its high operation potential and cycling performances. However, the low capacities and moisture uptake when exposed to air are the critical issues to be addressed. Among different anode materials, hard carbon is considered to be a better candidate for SIBs, due to its low working potentials and high cycling stability. Nevertheless, the sodium metal plating on the hard carbon surface at low voltage creates safety concerns because of the highly reactive nature of sodium metal. Moreover, the low capacity of ~300 mA h g^{-1} as compared to graphite led to the exploration of conversion and alloying reaction-based anode materials. The alloying and conversion type anode materials exhibited superior capacity, however, the high-volume expansion during the intercalation of larger sized Na^+ is the major drawback. Also, the low initial cycle Coulombic efficiency needs to be investigated. Many leading companies are promoting SIB research and are developing various prototypes of SIB for commercialization. For instance, Sumitomo Chemical Co. Ltd. fabricated the full-cell using the O3-type $NaNi_{0.3}Fe_{0.4}Mn_{0.3}O_2$ cathode, hard carbon anode, 1 M $NaPF_6$/PC electrolyte, and a polyethylene (PE) separator. The fabricated R2032 coin cell were demonstrated a capacity of 120 mAh g^{-1} normalized by the weight of the cathode active material.[32] Recently, framework of RS2E, a The French National Center for Scientific Research (CNRS) research network produces the first SIB battery in the 18650 format. The battery demonstrated an energy density of 90 Wh kg^{-1} with a cycling performance of 2000 cycles.[33] Toyota is working intensively on a prototype of a SIB for a car, the British company Faradion fabricated a battery pack for an electric bicycle.[34] The major restriction for the commercialization of a SIB is its low intrinsic energy density, and hence the low volumetric energy density which consumes a lot of space as compared to the LIBs. One way to mitigate this issue is by improving the tap density of the electrode materials. Another major concern is that in most cases the capacity and energy density of the SIB are estimated based on the electrode active material. However, the actual scenario in a commercial battery is different and thus while calculating these values, all the components that compose the SIB must be taken into consideration. Which leads to a further lower capacity and energy density value for the investigations SIBs to date. Thus, it is very important to improve the practical capacity and energy density of a SIB system, through the development of high performing cathode, anode, and stable electrolytes with appropriate additives. Further intensive research is required to modify the surface of the active materials to reduce the secondary reactions with electrolytes, binders, and other battery components. Thus, the existing understanding of the electrochemistry along with the new findings on materials and surfaces would be enough to accelerate the developments and commercialization of SIB technology in the future.

References

1. Kim, H., Hong, J., Park, K.-Y., Kim, H., Kim, S.-W., Kang, K. Aqueous rechargeable Li and Na ion batteries. *Chem. Rev.* **2014**, *114* (23), 11788.
2. Whitacre, J. F., Tevar, A., Sharma, S. Na$_4$Mn$_9$O$_{18}$ as a positive electrode material for an aqueous electrolyte sodium-ion energy storage device. *Electrochem. Commun.* **2010**, *12* (3), 463.
3. Kubota, K., Komaba, S. Review: Practical issues and future perspective for Na-ion batteries. *J. Electrochem. Soc.* **2015**, *162* (14), A2538.
4. Shacklette, L. W., Toth, J. E., Elsenbaumer, R. L. Conjugated polymer as substrate for the plating of alkali metal in a nonaqueous secondary batter. United States, **1987**, US Patent 4695521.
5. Kobayashi, Y., Shishikura, T., Konuma, H., Sakai, T., Nakamura, H., Takeuchi, M., Showa Denko Secondary battery. United States, **1985**, US Patent 4740436A.
6. Yabuuchi, N., Kajiyama, M., Iwatate, J., Nishikawa, H., Hitomi, S., Okuyama, R., Usui, R., Yamada, Y., Komaba, S. P2-type Na$_x$[Fe$_{1/2}$Mn$_{1/2}$]O$_2$ made from earth-abundant elements for rechargeable Na batteries. *Nat. Mat.* **2012**, *11*, 512.
7. Liao, Y., Park, K.-S., Xiao, P., Henkelman, G., Li, W., Goodenough, J. B. Sodium intercalation behavior of layered Na$_x$NbS$_2$ ($0 \leq x \leq 1$). *Chem. Mat.* **2013**, *25* (9), 1699.

8. Lu, Y., Wang, L., Cheng, J., Goodenough, J. B. Prussian blue: A new framework of electrode materials for sodium batteries. *Chem. Commun.* **2012**, *48* (52), 6544.
9. Zhao, R., Zhu, L., Cao, Y., Ai, X., Yang, H. X. An aniline-nitroaniline copolymer as a high capacity cathode for Na-ion batteries. *Electrochem. Commun.* **2012**, *21*, 36.
10. Su, D., Dou, S., Wang, G. Bismuth: A new anode for the Na-ion battery. *Nano Energy* **2015**, *12*, 88.
11. Yue, C., Yu, Y., Sun, S., He, X., Chen, B., Lin, W., Xu, B., Zheng, M., Wu, S., Li, J. et al. High performance 3D Si/Ge nanorods array anode buffered by TiN/Ti interlayer for sodium-ion batteries. *Adv. Funct. Mat.* **2015**, *25* (9), 1386.
12. Darwiche, A., Marino, C., Sougrati, M. T., Fraisse, B., Stievano, L., Monconduit, L. Correction to "Better cycling performances of bulk Sb in Na-ion batteries compared to Li-ion systems: An unexpected electrochemical mechanism". *J. Am. Chem. Soc.* **2013**, *135* (27), 10179.
13. Hu, Z., Wang, L., Zhang, K., Wang, J., Cheng, F., Tao, Z., Chen, J. MoS_2 nanoflowers with expanded interlayers as high-performance anodes for sodium-ion batteries. *Angewandte Chemie International Edition* **2014**, *53* (47), 12794.
14. Chevrier, V. L., Ceder, G. Challenges for Na-ion negative electrodes. *J. Electrochem. Soc.* **2011**, *158* (9), A1011.
15. Liu, J., Wen, Y., van Aken, P. A., Maier, J., Yu, Y. Facile synthesis of highly porous Ni–Sn intermetallic microcages with excellent electrochemical performance for lithium and sodium storage. *Nano Letters* **2014**, *14* (11), 6387.
16. Kim, Y., Park, Y., Choi, A., Choi, N.-S., Kim, J., Lee, J., Ryu, J. H., Oh, S. M., Lee, K. T. An amorphous red phosphorus/carbon composite as a promising anode material for sodium ion batteries. *Adv. Mat.* **2013**, *25* (22), 3045.
17. Qian, J., Wu, X., Cao, Y., Ai, X., Yang, H. High capacity and rate capability of amorphous phosphorus for sodium ion batteries. *Angewandte Chemie International Edition* **2013**, *52* (17), 4633.
18. Takeda, Y., Nakahara, K., Nishijima, M., Imanishi, N., Yamamoto, O., Takano, M., Kanno, R. Sodium deintercalation from sodium iron oxide. *Materials Research Bulletin* **1994**, *29* (6), 659.
19. Yoshida, H., Yabuuchi, N., Komaba, S. $NaFe_{0.5}Co_{0.5}O_2$ as high energy and power positive electrode for Na-ion batteries. *Electrochem. Commun.* **2013**, *34*, 60.
20. Kim, D., Lee, E., Slater, M., Lu, W., Rood, S., Johnson, C. S. Layered $Na[Ni_{1/3}Fe_{1/3}Mn_{1/3}]O_2$ cathodes for Na-ion battery application. *Electrochem. Commun.* **2012**, *18*, 66.
21. Stevens, D. A., Dahn, J. R. An in situ small-angle X-ray scattering study of sodium insertion into a nanoporous carbon anode material within an operating electrochemical cell. *J. Electrochem. Soc.* **2000**, *147* (12), 4428.
22. Stevens, D. A., Dahn, J. R. The mechanisms of lithium and sodium insertion in carbon materials. *J. Electrochem. Soc.* **2001**, *148* (8), A803.
23. Balogun, M.-S., Luo, Y., Qiu, W., Liu, P., Tong, Y. A review of carbon materials and their composites with alloy metals for sodium ion battery anodes. *Carbon* **2016**, *98*, 162.
24. Zhang, Y., Hou, H., Yang, X., Chen, J., Jing, M., Wu, Z., Jia, X., Ji, X. Sodium titanate cuboid as advanced anode material for sodium ion batteries. *J. Power Sources* **2016**, *305*, 200.
25. Doeff, M. M., Cabana, J., Shirpour, M. Titanate anodes for sodium ion batteries. *J. Inorg. Organomet. Polym. Mater.* **2014**, *24* (1), 5.

26. Cheng, Z., Mao, Y., Dong, Q., Jin, F., Shen, Y., Chen, L. Fluoroethylene carbonate as an additive for sodium-ion batteries: Effect on the sodium cathode. *Acta Phys Chim Sin.* **2019**, *35* (8), 868.
27. Komaba, S., Ishikawa, T., Yabuuchi, N., Murata, W., Ito, A., Ohsawa, Y. Fluorinated ethylene carbonate as electrolyte additive for rechargeable Na batteries. *ACS Appl. Mater. Interfaces* **2011**, *3* (11), 4165.
28. Shacklette, L. W., Jow, T. R., Townsend, L. Rechargeable electrodes from sodium cobalt bronzes. *J. Electrochem. Soc.* **1988**, *135* (11), 2669.
29. Yabuuchi, N., Komaba, S. Recent research progress on iron- and manganese-based positive electrode materials for rechargeable sodium batteries. *Sci. Technol. Adv. Mater.* **2014**, *15* (4), 043501.
30. Rahman, M.M., Xu, Y., Cheng, H., Shi,Q., Kou,R., Mu,L., Liu,Q. et al. Empowering multicomponent cathode materials for sodium ion batteries by exploring three-dimensional compositional heterogeneities. *Energy Environ Sci.* **2018**, *11* (9), 2496.
31. Chen, M., Hua, W., Xiao, J., Cortie, D., Chen, W., Wang, E., Hu, Z. et al. NASICON-type air-stable and all-climate cathode for sodium-ion batteries with low cost and high-power density. *Nat. Commun.* **2019**, *10* (1), 1480.
32. Kuze, S., Kageura, J.-I., Matsumoto, S., Nakayama, T., Makidera, M., Saka, M., Yamaguchi, T., Yamamoto, T., Nakane, K. Development of a sodium ion secondary battery. *SUMITOMO KAGAKU* **2013**, *2013*, 13.
33. Nemo, C. Na-ion batteries: a promising prototype! *Energy RS$_2$E,* **2015**, 1 https://www.energie-rs2e.com/en/news/na-ion-batteries-promising-prototype.
34. Edelstein, S. Faradion Electric Bike: Prototype Powered by Sodium-Ion Batteries. *Green car reports* **2015**, 1 https://www.greencarreports.com/news/1098434_faradion-electric-bike-prototype-powered-by-sodium-ion-batteries.

Index

Note: Page numbers in italic refer to figures.

A

acid, 9
active sites, 20, 105–106
alloying-based anode, 137
alloying reaction materials, 49–62
alluaudites, 105
aluminum substrate, 9, 19
amorphous carbon, 31
amorphous silicon, 50
anode, 2, 4–5, 23–65
  candidates, 4, 23, 37, 45, 50
anodic stability, 146
antimony, 54–60
arsenic, 54

B

battery, 10–18
  grade, 134
  packs, 10–11
battery management system (BMS), 11, 16
bilayered V_2O_5, 91
binder, 2, 137–139, *138*, *139*, 146–147
biomass-derived carbon, 31–32
bismuth, 54
black carbon, *40*

C

capacity, 16
carbon, 5, 24–32
  coating, 34, 43, 96
  doping, 108
  film, 30
  nanotubes, 36
carbonaceous materials, 23, 32
carboxylates functional groups, 63
cathode, 2, 3–4, 19, 83–115
cell, 9–10
cellulose, 137
charge/discharge, 12, 92, 111, 114
charged state, 14
charge storage mechanism, 8–9, 94
charging, 12, 14, 20
chemical energy, 20
chemical inertness, 132
clean energy, 7
cobalt oxides, 43
co-insertion, 134
commercial anode, 5, 23–24
conductive additions, 51
conversion reaction, 23, 41
copper, 9, 19
copper oxides, 44–45
Coulombic efficiency, 17–18, 32, *139*, 149
C-rate, 12, 18
crystalline silicon nanoparticles, 50
crystal structure, 51, 84, 100, 113–114
current collector, 6, 12, 41, 137
cut-off voltage, 16
cycle life, 16–17
cylindrical cell, 11, *11*

D

decomposition of electrolyte, 26, 138–139
defect position, 30
degradation, 4, 83
depth of discharge, 13
desodiation process, 3–4, 33–34, 41, 47, 50–51, 54, 60, 63, 134
discharging, 43, 132, 136
disordered graphene layers, 27
dissolved salts, 132, 134

E

electrical energy, 20
electrical insulating, 6, 132
electric vehicle, 7–8, 11, 86, 148
electrochemical mechanism, 8–9
electrochemical potential (voltage) window, *14*, 15–16

electrochemical processes, 9, 20
electrochemical reaction mechanism, 18–20, *19*
electrochemical stability, 6
electrodes, 9, 19–20, 144–146
electrolyte, 2, 5–6, 132–135, 146–147
　　additives, 136, 147
electromotive force, 13
energy, 18
　　energy density, 18
　　specific energy, 18
environmental friendly, 32–33
environmental pollution, 1
E-rate, 12
ether-typed electrolytes, 134
ethyl methanesulfonate-typed (EMS) organic electrolyte, 133
extraction, 91

F

facial fabrication process, 6, 32, 132
fluoroethylene carbonate (FEC), 135, 147
fluorophosphates, 101–103
fluoro-sulphate, 108
fossil fuels, 7
400 % volume changes, 56
functional groups, 30, 63–64

G

germanium, 51
graphene, 29–30
graphite, 24–26, *25*
grid, 1, 8, 143

H

hard carbon, 26–29
heteroatom doping, 30
hexacyanoferrates, 105–107
high capacity, 27, 30
high conductivity, 102
highest occupied molecular orbital (HOMO), 15–16
high-modulus, 137
high specific storage capacity, 1
high theoretical capacities, 45, 54, 56
high working potential, 37
hydrothermal technique, 93, 102

I

improved cycling stability, 58, 134
insertion, 23–41
　　reaction, 23
insertion/extraction process, 26–27, 85
intercalation/deintercalation process, 100, 104, 147
interconnection, 56
interlayer distance, 29–30, 58
internal resistance (IR), 13–15
　　drop, 15
introducing secondary phases, 91
ionically conductive, 132
ionic conductivity, 103, 108, 134
ionic radius, 2, 24
iron oxides, 42–43
irreversible capacity, 17
irreversible changes, 50

L

large volume changes, 6, 41, 54, 56, 137
layered sodium metal oxides, 84–90
　　O3 type material, 84–85
　　P2-type cathodes, 84
liquid electrolytes, 5–6
liquid organic solvents, 132
liquid solvents, 132
lithium compound, 1
lithium ion batteries (LIBs), 1–3, 7, 23, 83, 131, 143–145, 148–150
lithium titanate, 37–40
long cycling life, 1, 137
long time for drying, 132
low conductivity, 63
lowest unoccupied molecular orbital (LUMO), 15–16, 110
low working potential, 40–41

M

manganese oxides, 91–92
metal alloys, 23
metal oxides, 32–33; *see also* transition metal oxides
mixed polyanion, 4, 103–105
modules, 11
multidimensional, 114

Index

N

nanostructuring, 91, 98, 104
Na+ radius, 4, 83
negative electrode, 19, 144
nominal capacity, 16
nominal voltage, 16
non-polymeric materials, 109–111

O

olivine, 95–97
open-circuit voltage, 13–14
organic carbonyl materials, 63
organic compounds, 62–65, 109, 113–114
organic materials, 62, 108–112
organic solvents, 132, 136
overpotential, 15
oxidation, 15–16
oxygen evolution, 101, 146

P

petroleum coke, 26
phosphates, 95–105, 113
phosphorus, 54, 56–60
poly(acrylic acid), 137
polyanion compounds, 95
polymeric materials, 111–112
poly(vinylidene fluoride), 137
poor structural stability, 41, 56
porous structure, 32, 92
portable electronic devices, 1
positive electrode, 19
potassium ion batteries (KIBs), 2, 8
pouch cell, 11, *11*
power density, 18
    specific power, 18
prismatic cell, 11, *11*
promising anode candidate, 37, 44–45, 56
purity, 134
pyrophosphates, 100–101, 113

R

rate capacity, 4, 83
reagent grade, 134
rechargeable batteries, 18, 148
redox reaction, 103

reduced graphene oxide (rGO), 29–30, 94
reduced heat production, 134
reduced particle size, 34, 43
reduction processes, 9, 15, 24, 132
renewable and cleaner energy sources, 1
reversible capacity, 23–24, 26–27, 29–30, 32, 37, 39, 45, 47–48, 50–51, 58, 62, 64, 89, 107
reversible redox reaction, 51

S

safety concerns, 149
scalable production, 6, 132
SEI film, 5–6, 43, 131, 134–139
self-discharge, 17
separator, 10
silicon, 50–51
sodiation process, 3, 24, 33, 37, 41, 44, 47, 50, 60, 62
sodium alginate, 137, 147
sodium-based mixed-cation oxides, 88–90
sodium carboxymethyl cellulose (Na-CMC), 137
sodium-free metal oxides, 90–93
sodium ion batteries (SIBs), 2, 2, 8, 83, 131
sodium salts, 132, 146
sodium super ionic conductor (NASICON), 98–99
soft carbon, 27
solar, 1
sol–gel process, 92, 98–99
solid electrolyte interfaces (SEI), 5–6, 43, 131, 134–139, 144
solid electrolytes, 98
solid-phase method, 92
solid-state electrolytes, 9, 113
solvothermal technique, 99, 103
stable SEI films, 6, 43, 135, 137
state of charge, 12
stripping, 6, 137
strong chemical bonding, 6, 137
structural degradation, 86
substrate, 9, 19
suitable working potential, 1
sulfates, 107–108
superior conductivity, 91

T

tavorite, 107
terminal voltage, 13
theoretical capacity, 47, 51, 100
thermal runaway, 146
thermal stability, 14, 99, 134, 145
3D framework, 137
tight contact, 6, 137
tin, 51–54
tin oxide, 44
titanium, 32–33
titanium oxides, 33–37
toxic, 47, 137, 147
transition metal, 4, 41
    fluorides, 93–95
    oxides, 42–45, 84–93
    phosphides, 5, 41, 48
    sulfide, 5, 41, 45–48
transparent liquid, 134
trickle charge, 14
trilayers, 47
tunnel type sodium transition-metal oxides, 92–93
two-dimensional (2D) carbon material, 29

V

vanadium oxides, 90–91
vanadyl phosphate, 100
void space, 48, 51, 62
voltage hysteresis, 15
voltage polarization, 15
volume changes, 4, 6, 41, 50, 54, 56, 103–104, 136–137
volume expansion, 20, 43–45, 51, 55–56, 83, 90

W

wave, 1
wind, 1, 7
wood, 32
working potential, 34, 37

Y

yolk-shell structure, 51